Edited by Harendra Singh, Devendra Kumar and Dumitru Baleanu

Methods of Mathematical Modelling

Mathematics and Its Applications: Modelling, Engineering, and Social Sciences

Series Editor: Hemen Dutta
Department of Mathematics, Gauhati University

Tensor Calculus and Applications:
Simplified Tools and Techniques
Bhaben Kalita

Concise Introduction to Logic and Set Theory
Iqbal H. Jebril and Hemen Dutta

Discrete Mathematical Structures:
A Succinct Foundation
Beri Venkatachalapathy Senthil Kumar and Hemen Dutta

Methods of Mathematical Modelling:
Fractional Differential Equations
Edited by Harendra Singh, Devendra Kumar, and Dumitru Baleanu

For more information about this series, please visit:
https://www.crcpress.com/Mathematics-and-its-applications/book-series/MES

Edited by Harendra Singh, Devendra Kumar and Dumitru Baleanu

Methods of Mathematical Modelling

Fractional Differential Equations

CRC Press is an imprint of the
Taylor & Francis Group, an **informa** business

CRC Press
Taylor & Francis Group
6000 Broken Sound Parkway NW, Suite 300
Boca Raton, FL 33487-2742

First issued in paperback 2020

ISBN-13: 978-0-367-22008-2 (hbk)
ISBN-13: 978-0-367-77655-8 (pbk)

Library of Congress Cataloging-in-Publication Data

Names: Singh, Harendra, editor. | Kumar, Devendra, editor. | Baleanu, D. (Dumitru), editor.
Title: Methods of mathematical modelling : fractional differential equations / edited by Harendra Singh, Devendra Kumar, and Dumitru Baleanu.
Other titles: Methods of mathematical modelling (CRC Press)
Description: Boca Raton, FL : CRC Press/Taylor & Francis Group, 2019. | Summary: "Mathematical modelling is a process which converts real-life problems into mathematical problems whose solutions make it easy to understand the real-life problem. Fractional modeling has many real-life applications in mathematics, science, and engineering. Such as viscoelasticity, chemical engineering, signal processing, bioengineering, control theory, and fluid mechanics. This book offers a collection of chapters on classical and modern dynamical systems modelled by fractional differential equations. This book will be useful to readers in increasing their knowledge in this field" – Provided by publisher.
Identifiers: LCCN 2019020280 | ISBN 9780367220082 (hardback : acid-free paper) | ISBN 9780429274114 (ebook)
Subjects: LCSH: Mathematical models. | Fractional differential equations.
Classification: LCC TA342 .M43 2019 | DDC 515/.35–dc23
LC record available at https://lccn.loc.gov/2019020280

Visit the Taylor & Francis Web site at
http://www.taylorandfrancis.com

and the CRC Press Web site at
http://www.crcpress.com

Contents

Preface

This book is planned for graduate students and researchers working in the area of mathematical modelling and fractional calculus. It describes several useful topics in mathematical modelling having real-life applications in chaos, physics, fluid mechanics and chemistry. The book consists of thirteen chapters and is organized as follows:

Chapter 1 presents the dynamical behaviour of two ecosystems of three species consisting of prey, intermediate predator and top-predator that are still of current and recurring interests. The classical integer-order derivatives in such models are replaced with the Atangana–Baleanu fractional derivative in the sense of Caputo. Existence and uniqueness of solution are established. Linear stability analysis is examined in a view to guide the correct choice of parameters when numerically simulating the models. In the analysis, the condition for a dynamic system to be locally asymptotically stable is provided. A range of chaotic and spatiotemporal phenomena are obtained for different instances of $\alpha \in (0, 1)$ and are also given to justify the theoretical findings.

Chapter 2 investigates the solutions for fractional diffusion equations subjected to reactive boundary conditions. For this, the system is defined in a semi-infinity medium, and the presence of a surface that may adsorb, desorb and/or absorb particles from the bulk is considered. The particles absorbed from the bulk by the surface may promote, by a reaction process, the formation of other particles. The particle dynamics is governed by generalized diffusion equations in the bulk and by kinetic equations on the surface; consequently, memory effects are taken into account in order to enable an anomalous diffusion approach and, consequently, non-Debye relaxations. The results exhibit a rich variety of behaviour for the particles, depending on the choice of characteristic times present in the boundary conditions or the fractional index present in modelling equations.

Chapter 3 presents an efficient computational method for the approximate solution of the non-linear fractional Lienard equation (FLE), which describes the oscillating circuit. The Lienard equation is a generalization of the spring-mass system equation. The fractional derivative is in a Liouville–Caputo sense. The computational method is a combination of collocation method and operational matrix method for Legendre scaling functions. Behaviour of solutions for different fractional order is shown through figures.

Chapter 4 introduces a new approximation scheme to solve fractional differential equations with Gomez–Atangana–Caputo derivatives. The algorithm is easy to use and converges very quickly. Some examples are presented with some numerical simulations to show the efficiency of the proposed scheme. Comparisons are made with exact solutions or numerical solutions obtained with other methods.

Chapter 5 presents a spectral formulation for a fractional optimal control problem (FOCP) defined in spherical coordinates in the cases of half and complete axial symmetry. The dynamics of an optimally controlled system are described by space–time fractional diffusion equation in terms of the Caputo and fractional Laplacian differentiation operators. The first step in numerical methodology is to represent the state and control functions of the system as eigenfunction expansion series. This is clearly obtained by the discretization of space fractional Laplacian operator term. In the next step, the necessary optimality conditions are determined by using fractional Euler–Lagrange equations. Therefore, the space–time fractional differential equation is reduced into time fractional differential equations in terms of forward and backward fractional Caputo derivatives. In the last step, the time domain is discretized into a number of subintervals by using the Grunwald–Letnikov approach. An illustrative example is considered for various orders of fractional derivatives and different spatial and temporal discretizations. As a result, limited number of grid points is sufficient to obtain good results. In addition, the numerical solutions converge as the size of the time step is reduced. Note that the solution techniques in the sense of analytical or numerical for space–time fractional differential equations defined by fractional Laplacian, Riesz or Riesz–Feller spatial derivatives are quite complicated. In reality, these equations correspond to systems that show the behaviour of anomalous diffusion. For this complexity, the spectral approaches depend on the spatial domain on which the main problem is constructed, are clearer and so has extensive applicability than the alternative ones.

Chapter 6 presents a new approximate solution for the fractional diffusion equation described by the Caputo-generalized fractional derivative. The heat balance integral method and the double integral method have been used for getting an approximate solution of the generalized fractional diffusion equation. The effect of the order ρ in the diffusion process has been analyzed. The approximate solution of the fractional diffusion equation described by the integer order derivative, the Caputo derivative and Caputo-generalized fractional derivative have been proposed and compared.

Chapter 7 presents an analytical algorithm for non-linear time fractional Toda lattice equations. The proposed algorithm is a new amalgamation of the homotopy analysis, Laplace transform and Adomian's polynomials. First, an alternative framework of the proposed method, which can be simply used to effectively handle non-linear problems arising in science and engineering, is presented. Comparisons are made between the results of the proposed method and exact solutions. An illustrative example is given to demonstrate the simplicity and efficiency of the proposed amalgamation of the homotopy analysis, Laplace transform and Adomian's polynomials. The results reveal that the proposed method is explicit, efficient and easy to use. The fractional derivatives here are described in a Caputo sense.

Chapter 8 applies the idea of the fractional derivative to the heat transfer problem of a hybrid nanofluid. More exactly, this chapter deals with the generalization of natural convection flow of $Cu - Al_2O_3 - H_2O$ hybrid nanofluid in two infinite vertical parallel plates. To demonstrate the flow phenomena in two parallel plates of hybrid nanofluid, the Brinkman-type fluid model is utilized. The governing equation of Brinkman-type fluid together with the energy equation is subjected to appropriate initial and boundary conditions. The Caputo–Fabrizio fractional derivative approach is used for the generalization of the mathematical model. The Laplace transform technique is used to develop exact analytical solutions for velocity and temperature profiles. The general solutions for velocity and temperature profiles are brought into light through numerical computation and graphical representation. The obtained results show that the velocity and temperature profiles show dual behaviour for $0 < \alpha < 1$ and $0 < \beta < 1$, where α and β are the fractional parameters. It is noticed that, for a shorter time, the velocity and temperature distributions decrease with an increase in the values of fractional parameters, whereas the trend reverses for an increase in time.

Chapter 9 solves mathematical model obtained due to groundwater and is recharged by rain water or spreading the water on the ground in vertical direction; hence, the wetness of the soil increases using the Caputo, Caputo–Fabrizio and Atangana–Baleanu fractional derivative operators, respectively. The q-homotopy analysis method is applied to obtain the solutions of the equations and is affirmed by comparing the model results with those available in the literature. This method has great freedom to choose the auxiliary parameter h, auxiliary function $H(\xi, T)$ and the initial guesses. The convergence region of solution series can be adjusted and controlled by choosing proper values for auxiliary parameter h and auxiliary function $H(\xi, T)$.

Chapter 10 studies a range of chaotic and hyperchaotic processes modelled with the Atangana–Baleanu fractional derivative that has both non-local and non-singular properties in the sense of Caputo. A modified Chua chaotic attractor have been extended and analyzed within the scope of fractional differentiation and integration. Three cases of fractional differential operators are considered, namely the Caputo, Caputo–Fabrizio and the Atangana–Baleanu derivatives. Fixed point theory and approximation method are applied to show the existence and uniqueness of solutions. Due to non-linearity of this modified model, a user-friendly scheme is used to provide numerical solutions.

Chapter 11 discusses a new numerical method, namely the Adomian decomposition Sumudu transform method (ADSTM), to find the numerical solution of non-linear time-fractional Zakharov–Kuznetsov (FZK) equation in two dimensions. The suggested technique is applied on two test examples and the solution is graphically presented.

Chapter 12 studies the propagation of wave envelop with fractional temporal evolution by considering the transverse surface in the non-linear dynamic peer-reviewed international journals. His 121 research papers have been published in various journals of repute with an h-index of 26. He has attended a number of national and international conferences and presented several research papers. He has also attended summer courses, short-term programs and workshops. He is a member of the Editorial Board of various journals of mathematics and is a reviewer of various journals.

Dumitru Baleanu is a professor in the Department of Mathematics, Cankaya University, Ankara, Turkey, and Institute of Space Sciences, Magurele-Bucharest, Romania. His research interests include fractional dynamics and its applications, fractional differential equations, dynamic systems on time scales, Hamilton-Jacobi formalism and Lie symmetries. He has published more than 600 papers indexed in SCI. He is one of the editors of five books published by Springer and one published by AIP *as AIP Conference Proceedings* and a co-author of the monograph book titled *Fractional Calculus: Models and Numerical Methods*, published in 2012 by World Scientific Publishing. He is an editorial board member for the following journals indexed in SCI: *Mathematics*, *Journal of Vibration and Control*, *Symmetry*, *Frontiers in Physics*, *Open Physics*, *Advances in Difference Equations* and *Fractional Calculus and Applied Analysis*. He is the editor in chief of the *Progress of Fractional Differentiation and Applications*. He is also a member of the Editorial Board of 12 different journals that are not indexed in SCI. He is a member of the advisory board of the 'Mathematical Methods and Modeling for Complex Phenomena' book collection, published jointly by Higher Education Press and Springer. Also, he was a scientific board member for the Chemistry and Physics of InTech Scientific Board in 2011/2012. He has received more than 8,000 citations (excluded from citation overview: self-citations of all authors) in journals covered by SCI, with an h-index of 46. He was on the Thompson Reuter list of highly cited researchers in 2015, 2016, 2017 and 2018.

Contributors

Ritu Agarwal
Department of Mathematics
Malaviya National Institute of Technology
Jaipur, Rajasthan, India

Ravi P. Agarwal
Department of Mathematics
Texas A&M University—Kingsville
Kingsville, Texas

Abdon Atangana
Département de Mathématiques de la Décision
Université Cheikh Anta Diop de Dakar
Dakar Fann, Senegal
and
Institute for Groundwater Studies
University of the Free State
Bloemfontein, South Africa

Derya Avci
Department of Mathematics
Balikesir University
Balikesir, Turkey

Dépélair Bienvenue
Department of Physics
The University of Maroua
Maroua, Cameroon

Betchewe Gambo
Department of Physics
The University of Maroua
Maroua, Cameroon

Zakia Hammouch
Department of Mathematics
FST Errachidia, Moulay Ismail University of Meknes
Errachidia, Morocco

Alphonse Houwe
Department of Physics
The University of Maroua
Maroua, Cameroon
and
Department of Marine Engineering
Limbe Nautical Arts and Fisheries Institute
Limbe, Cameroon

K. Jothimani
Department of Mathematics
Sri Eshwar College of Engineering
Coimbatore, Tamil Nadu, India

Berat Karaagac
Department of Mathematics Education
Adyaman University
Adyaman, Turkey

Hardish Kaur
Department of Mathematics
National Institute of Technology
Kurukshetra, Haryana, India

Ilyas Khan
Department of Mathematics
College of Science Al-Zulfi, Majmaah University
Al-Majmaah, Saudi Arabia

Amit Kumar
Department of Mathematics
Balarampur College
Purulia, West Bengal, India

Devendra Kumar
Department of Mathematics
University of Rajasthan
Jaipur, Rajasthan, India

Ranbir Kumar
Department of Mathematics
National Institute of Technology
Jamshedpur, Jharkhand, India

Sunil Kumar
Department of Mathematics
National Institute of Technology
Jamshedpur, Jharkhand, India

Ervin K. Lenzi
Departamento de Física
Universidade Estadual de
Ponta Grossa, Brazil

Marcelo K. Lenzi
Departamento de Engenharia Química
Universidade Federal do Paraná
Curitiba, Brazil

Toufik Mekkaoui
Department of Mathematics
Faculty of Sciences and Techniques, Moulay Ismail University of Meknes
Errachidia, Morocco

Justin Mibaille
Higher Teachers' Training College of Maroua
The University of Maroua
Maroua, Cameroon

Jyoti Mishra
Department of Mathematics
Gyan Ganga Institute of Technology and Sciences
Jabalpur, Madhya Pradesh, India

Kolade M. Owolabi
Institute for Groundwater Studies
University of the Free State
Bloemfontein, South Africa
and
Department of Mathematical Sciences
Federal University of Technology
Akure, Nigeria

Necati Ozdemir
Department of Mathematics
Balikesir University
Balikesir, Turkey

Amit Prakash
Department of Mathematics
National Institute of Technology
Kurukshetra-136119, Haryana, India

C. Ravichandran
PG and Research Department of Mathematics
Kongunadu Arts and Science College (Autonomous)
Coimbatore, Tamil Nadu, India

Muhammad Saqib
Department of Mathematical Sciences
Universiti Teknologi Malaysia JB
Skudai, Malaysia

Ndolane Sene
Département de Mathématiques de la Décision
Université Cheikh Anta Diop de Dakar
Dakar Fann, Senegal
and
Institute for Groundwater Studies
University of the Free State
Bloemfontein, South Africa

Sharidan Shafie
Department of Mathematical Sciences
Universiti Teknologi Malaysia
Skudai, Malaysia

Harendra Singh
Department of Mathematics
Post-Graduate College
Ghazipur, Uttar Pradesh, India

Jagdev Singh
Department of Mathematics
JECRC University
Jaipur, Rajasthan, India

Mahaveer Prasad Yadav
Department of Mathematics
Malaviya National Institute of Technology
Jaipur, Rajasthan, India

Mehmet Yavuz
Department of Mathematics-Computer
Necmettin Erbakan University
Konya, Turkey

1

Mathematical Analysis and Simulation of Chaotic Tritrophic Ecosystem Using Fractional Derivatives with Mittag-Leffler Kernel

Kolade M. Owolabi
University of the Free State
Federal University of Technology

Berat Karaagac
Adyaman University

CONTENTS

1.1 Introduction

In the past few decades, population systems consisting of one or two species have attracted the attention of scientists and other scholars [2,10,15–17,25,28]. It was observed that only a few handful of research findings reported on multi-component systems of three or more species [20]. The aim of this work is to study the dynamics of predator–prey model consisting of spatial interactions among the prey, intermediate predator and the top predator. The concept of predation could be dated back to the pioneering predator–prey system of Lotka and Volterra in 1925 and 1926, respectively [12, 29].

Many research scholars have reported that modelling with fractional calculus concept is very suitable and reliable to give an accurate description of memory and some physical properties of various materials and processes, which are completely missing in classical or integer-order equations, and a fractional mathematical model can give more reliable information about real-life phenomena [3–5, 27]. In addition, many physical systems encountered in various disciplines have been described by fractional differential equations, which include the hydrology and groundwater flow [4, 7], diffusion-like waves, pattern formation in chemical and biological processes [9, 18, 26], non-linear movement of earthquakes [9], viscoelastic materials [14] and muscular blood vessel model [1], among several other applications. Also, fractional-order problems are naturally connected to models with memory, which arise in some biological scenarios [22–24].

Hence, we are motivated by the dynamics of multi-species ecosystems with fractional derivative in this work. Some of the definitions and properties of fractional derivatives are as follows.

The Riemann–Liouville fractional derivative of order $\alpha \in (0, 1]$ for a function $u(t) \in C^1([0, b], \mathbb{R}^n)$; $b > 0$ is given by [27]

$$ {}^{RL}\mathcal{D}_0^{\alpha}u(t) = \frac{d^n}{dt^n}\mathcal{I}_0^{1-\alpha}u(t) = \frac{1}{\Gamma(1-\alpha)}\frac{d^n}{dt^n}\int_0^t (t-\xi)^{n-\alpha-1}u(\xi)d\xi, \quad (1.1) $$

for all $t \in [0, b]$ and $n - 1 < \alpha < n$, where $n > 0$ is an integer.

The Caputo fractional derivative of order $\alpha \in (0, 1]$ for a function $u(t) \in C^1([0, b], \mathbb{R}^n)$; $b > 0$ is given by [27]

$$ {}^{C}\mathcal{D}_{0,t}^{\alpha}u(t) = \frac{1}{\Gamma(n-\alpha)}\int_0^t (t-\xi)^{n-\alpha-1}u^n(\xi)d\xi, \quad (1.2) $$

for all $t \in [0, b]$.

Recently, Atangana and Baleanu proposed a new fractional derivative with non-local and non-singular kernels in the sense of Caputo as [3]

$$ {}^{ABC}\mathbf{D}_t^{\alpha}[u(t)] = \frac{M(\alpha)}{1-\alpha}\int_0^t u'(\xi)E_{\alpha}\left[-\alpha\frac{(t-\xi)^{\alpha}}{1-\alpha}\right]d\xi \quad (1.3) $$

where $M(\alpha)$ has the same definition as in the case of the Caputo–Fabrizio fractional derivative [8], and E_{α} is a one-parameter Mittag-Leffler function given as

$$ u(z) = E_{\alpha}(z) = \sum_{k=0}^{\infty}\frac{z^k}{\Gamma(\alpha k+1)}, \quad \alpha > 0, \ \alpha \in \mathbb{R}, \ z \in \mathbb{C}. \quad (1.4) $$

The derivative given in (1.3) which we are applying in the present work is popularly called the Atangana–Baleanu fractional operator in the sense of Caputo. This derivative has been applied to model a number of real-life phenomena. The remainder part of this work is arranged as follows. The method of approximation of fractional derivative is given in Section 1.2. Model equations and analysis are introduced in Section 1.3. Numerical experiments that confirm the analytical findings are reported for some instances of fractional power in Section 1.4. The conclusion is drawn in the last section.

1.2 Method of Approximation of Fractional Derivative

In this section, we follow closely the approximation techniques reported in [6] to approximate the novel Atangana–Baleanu fractional operator as follows.

Consider the differential equation

$$ {}_0^{ABC}\mathcal{D}_t^{\alpha} z(t) = F(t, z(t)), \tag{1.5}$$

where $z = (u, v, w)$ and $F = f_1(u, v, w), f_2(u, v, w), f_3(u, v, w)$. Follow [6] and apply the fundamental calculus theory to have

$$ z(t) - z(0) = \frac{1-\alpha}{ABC(\alpha)} F(t, z(t)) + \frac{\alpha}{ABC(\alpha)\Gamma(\alpha)} \int_0^t (t-\tau)^{\alpha-1} F(\tau, z(\tau)) d\tau. \tag{1.6}$$

At t_{n+1}, we have

$$\begin{aligned} z(t_{n+1}) - z(0) &= \frac{1-\alpha}{ABC(\alpha)} F(t_n, z_n) \\ &+ \frac{\alpha}{ABC(\alpha)\Gamma(\alpha)} \int_0^{t_{n+1}} (t_{n+1} - \tau)^{\alpha-1} F(t, z(t)) dt \end{aligned}$$

and at t_n leads to

$$\begin{aligned} z(t_n) - z(0) &= \frac{1-\alpha}{ABC(\alpha)} f(t_{n-1}, z_{n-1}) \\ &+ \frac{\alpha}{ABC(\alpha)\Gamma(\alpha)} \int_0^{t_n} (t_n - \tau)^{\alpha-1} F(t, z(t)) dt \end{aligned}$$

on subtraction leads to

$$\begin{aligned} z(t_{n+1}) - z(t_n) &= \frac{1-\alpha}{ABC(\alpha)} \{F(t_n, z_n) - F(t_{n-1}, z_{n-1})\} \\ &+ \frac{\alpha}{ABC(\alpha)\Gamma(\alpha)} \int_0^{t_{n+1}} (t_{n+1} - t)^{\alpha-1} F(t, z(t)) dt \\ &- \frac{\alpha}{ABC(\alpha)\Gamma(\alpha)} \int_0^{t_n} (t_n - t)^{\alpha-1} F(t, z(t)) dt. \end{aligned} \tag{1.7}$$

Thus,

$$z(t_{n+1}) - z(t_n) = \frac{1-\alpha}{ABC(\alpha)}\{F(t_n, z_n) - F(t_{n-1}, z_{n-1})\} + E_{\alpha,1} - E_{\alpha,2}.$$

Next, we consider

$$E_{\alpha,1} = \frac{\alpha}{ABC(\alpha)\Gamma(\alpha)}\int_0^{t_{n+1}} (t_{n+1}-t)^{\alpha-1} F(t, z(t))dt$$

With approximation

$$q(t) = \frac{t - t_{n-1}}{t_n - t_{n-1}} F(t_n, z_n) + \frac{t - t_{n-1}}{t_{n-1} - t_n} F(t_{n-1}, z_{n-1}) \tag{1.8}$$

we get

$$\begin{aligned}
E_{\alpha,1} &= \frac{\alpha}{ABC(\alpha)\Gamma(\alpha)}\int_0^{t_{n+1}} (t_{n+1}-t)^{\alpha-1} \\
&\quad \times \left\{\frac{t - t_{n-1}}{h} F(t_n, z_n) - \frac{t - t_n}{h} F(t_n, z_n)\right\} \\
&= \frac{\alpha F(t_n, z_n)}{ABC(\alpha)\Gamma(\alpha)h}\left\{\int_0^{t_{n+1}} (t_{n+1}-t)^{\alpha-1} F(t - t_{n-1})\right\} dt \\
&\quad - \frac{\alpha F(t_{n-1}, z_{n-1})}{ABC(\alpha)\Gamma(\alpha)h}\left\{\int_0^{t_{n+1}} (t_{n+1}-t)^{\alpha-1} F(t - t_{n-1})\right\} dt \\
&= \frac{\alpha F(t_n, z_n)}{ABC(\alpha)\Gamma(\alpha)h}\left\{\frac{2ht_{n+1}^{\alpha}}{\alpha} - \frac{t_{n+1}^{\alpha+1}}{\alpha+1}\right\} \\
&\quad - \frac{\alpha F(t_{n-1}, z_{n-1})}{ABC(\alpha)\Gamma(\alpha)h}\left\{\frac{ht_{n+1}^{\alpha}}{\alpha} - \frac{t_{n+1}^{\alpha+1}}{\alpha+1}\right\},
\end{aligned} \tag{1.9}$$

and

$$E_{\alpha,2} = \frac{\alpha F(t_n, z_n)}{ABC(\alpha)\Gamma(\alpha)h}\left\{\frac{ht_n^{\alpha}}{\alpha} - \frac{t_n^{\alpha+1}}{\alpha+1}\right\} - \frac{F(t_{n-1}, z_{n-1})}{ABC(\alpha)\Gamma(\alpha)h}. \tag{1.10}$$

Thus,

$$\begin{aligned}
z(t_{n+1}) - z(t_n) &= \frac{1-\alpha}{ABC(\alpha)}\{F(t_n, z_n) - F(t_{n-1}, z_{n-1})\} + \frac{\alpha F(t_n, z_n)}{ABC(\alpha)\Gamma(\alpha)h} \\
&\quad \times \left\{\frac{2ht_{n+1}^{\alpha}}{\alpha} - \frac{t_{n+1}^{\alpha+1}}{\alpha+1}\right\} - \frac{\alpha F(t_{n-1}, z_{n-1})}{ABC(\alpha)\Gamma(\alpha)h}\left\{\frac{ht_{n+1}^{\alpha}}{\alpha} - \frac{t_{n+1}^{\alpha+1}}{\alpha+1}\right\} \\
&\quad - \frac{\alpha F(t_n, z_n)}{ABC(\alpha)\Gamma(\alpha)h}\left\{\frac{ht_n^{\alpha}}{\alpha} - \frac{t_n^{\alpha+1}}{\alpha+1}\right\} + \frac{F(t_{n-1}, z_{n-1})}{ABC(\alpha)\Gamma(\alpha)} t_n^{\alpha+1}
\end{aligned} \tag{1.11}$$

$$z_{n+1} = z_n + F(t_n, z_n)\left\{\frac{1-\alpha}{ABC(\alpha)} + \frac{\alpha}{ABC(\alpha)h}\left[\frac{2ht_{n+1}^{\alpha}}{\alpha} - \frac{t_{n+1}^{\alpha+1}}{\alpha+1}\right]\right.$$
$$\left. - \frac{\alpha}{ABC(\alpha)\Gamma(\alpha)h}\left[\frac{ht_n^{\alpha}}{\alpha} - \frac{t_n^{\alpha+1}}{\alpha+1}\right]\right\} + F(t_{n-1}, z_{n-1})$$
$$\times \left\{\frac{\alpha-1}{ABC(\alpha)} - \frac{\alpha}{h\Gamma(\alpha)ABC(\alpha)}\right.$$
$$\left.\times \left[\frac{ht_{n+1}^{\alpha}}{\alpha} - \frac{t_{n+1}^{\alpha+1}}{\alpha+1} + \frac{t^{\alpha+1}}{h\Gamma(\alpha)ABC(\alpha)}\right]\right\}. \tag{1.12}$$

Scheme (1.12) is known as the two-step Adams–Bashforth method for the approximation of Atangana–Baleanu fractional derivative.

Existence and uniqueness of the solution for a general three-species system is briefly discussed via the fractional derivative operator. The general multi-species fractional system of ordinary differential equation in the sense of Atangana–Baleanu derivative of order α is given in compact form

$${}_0^{ABC}\mathcal{D}_t^{\alpha}\mathbf{U}(t) = F(\mathbf{U}, t), \quad \mathbf{U}(0) = \mathbf{U}^0, t \in (0, T], \tag{1.13}$$

where

$$\mathbf{U} = \begin{pmatrix} u^1 \\ u^2 \\ \vdots \\ u^n \end{pmatrix}, \quad \mathbf{U}^0 = \begin{pmatrix} u_0^1 \\ u_0^2 \\ \vdots \\ u_0^n \end{pmatrix}, \quad F(\mathbf{U}, t) = \begin{pmatrix} f_1(u_1, u_2, \ldots, u_n, t) \\ f_1(u_1, u_2, \ldots, u_n, t) \\ \vdots \\ f_n(u_1, u_2, \ldots, u_n, t). \end{pmatrix}$$

The supremum norm is defined as

$$\|\mathbf{S}\| = \sup_{t\in(0,T]} |\mathbf{S}|,$$

we define the norm of matrix $\mathbf{A} = [a_{ij}[t]]$ by

$$\|\mathbf{A}\| = \sum_{i,j} \sup_{t\in(0,T]} |a_{ij}[t]|.$$

So for three-species system, we have

$$\begin{aligned} {}_0^{ABC}\mathcal{D}_t^{\alpha}u(t) &= f_1(u, v, w, t), \\ {}_0^{ABC}\mathcal{D}_t^{\alpha}v(t) &= f_2(u, v, w, t), \\ {}_0^{ABC}\mathcal{D}_t^{\alpha}w(t) &= f_3(u, v, w, t). \end{aligned} \tag{1.14}$$

By applying the fundamental calculus theorem to the components [22], one obtains

$$\begin{aligned}
u(t)-u(0)&=\frac{1-\alpha}{AB(\alpha)}f_1(u,v,w,t)+\frac{\alpha}{AB(\alpha)\Gamma(\alpha)}\int_0^t(t-\tau)^{\alpha-1}f_1(u,v,w,\tau)d\tau\\
v(t)-v(0)&=\frac{1-\alpha}{AB(\alpha)}f_2(u,v,w,t)+\frac{\alpha}{AB(\alpha)\Gamma(\alpha)}\int_0^t(t-\tau)^{\alpha-1}f_2(u,v,w,\tau)d\tau\\
w(t)-w(0)&=\frac{1-\alpha}{AB(\alpha)}f_3(u,v,w,t)+\frac{\alpha}{AB(\alpha)\Gamma(\alpha)}\int_0^t(t-\tau)^{\alpha-1}f_3(u,v,w,\tau)d\tau
\end{aligned} \tag{1.15}$$

We require $\mathscr{P}_{a,b}$, a compact in the form

$$\mathscr{P}_{a,b}=I_a(t_0)\times\mathcal{B}_b(\zeta) \tag{1.16}$$

where

$$\zeta=\min\{u_0,u_0\},$$

with similar expression for variables v, w, and

$$I_a(t_0)=[t_0-a,t_0+a],\quad \mathcal{B}_0(\zeta)=[\xi-b,\tau+b].$$

Let

$$S=\max_{\mathscr{P}_{a,b}}\left\{\sup_{\mathscr{P}_{a,b}}\|f_1\|,\ \sup_{\mathscr{P}_{a,b}}\|f_2\|,\ \sup_{\mathscr{P}_{a,b}}\|f_3\|\right\}.$$

By adopting infinite norm yields

$$\|\Phi\|_\infty=\sup_{t\in I_a}||\Phi(t)|.$$

Next, we create a function, say

$$\Gamma:\mathscr{P}_{a,b}\to\mathscr{P}_{a,b}$$

in such a way that

$$\begin{aligned}
\Gamma u(t)&=u_0+\frac{1-\alpha}{AB(\alpha)}f_1(u,v,w,t)+\frac{\alpha}{AB(\alpha)\Gamma(\alpha)}\int_0^t f_1(u,v,w,t)(t-\tau)^{\alpha-1}d\tau\\
\Gamma v(t)&=v_0+\frac{1-\alpha}{AB(\alpha)}f_2(u,v,w,t)+\frac{\alpha}{AB(\alpha)\Gamma(\alpha)}\int_0^t f_2(u,v,w,t)(t-\tau)^{\alpha-1}d\tau\\
\Gamma w(t)&=w_0+\frac{1-\alpha}{AB(\alpha)}f_3(u,v,w,t)+\frac{\alpha}{AB(\alpha)\Gamma(\alpha)}\int_0^t f_3(u,v,w,t)(t-\tau)^{\alpha-1}d\tau
\end{aligned} \tag{1.17}$$

Next, we evaluate the following conditions to show that the fractional derivative operator is well defined

$$\begin{aligned}
&\|\Gamma_1u(t)-u_0\|_\infty<b,\\
&\|\Gamma_2v(t)-v_0\|_\infty<b,\\
&\|\Gamma_2w(t)-w_0\|_\infty<b.
\end{aligned} \tag{1.18}$$

So, beginning with species u, we have

$$\begin{aligned}
\|\Gamma_1 u(t) - u_0\|_\infty &= \left\| \frac{1-\alpha}{AB(\alpha)} f_1(u,v,w,t) + \frac{\alpha}{AB(\alpha)\Gamma(\alpha)} \right. \\
&\quad \left. \times \int_0^t f_1(u,v,w,t)(t-\tau)^{\alpha-1} d\tau \right\|_\infty \\
&\leq \frac{1-\alpha}{AB(\alpha)} \|f_1(u,v,w,t)\|_\infty + \frac{\alpha}{AB(\alpha)\Gamma(\alpha)} \|f_1(u,v,w,t) \\
&\quad \times \int_0^t (t-\xi) d\xi \\
&\leq \frac{(1-\alpha)M}{AB(\alpha)} + \frac{\alpha M}{AB(\alpha)\Gamma(\alpha)} \cdot a^\alpha < b \qquad (1.19)
\end{aligned}$$

This means that

$$a = \left(\frac{b - \frac{(1-\alpha)M}{AB(\alpha)}}{\frac{\alpha M}{AB(\alpha)\Gamma(\alpha)}} \right)^{\frac{1}{\alpha}}.$$

Also, it is required need to show that functions $u(t)$, $v(t)$ and $w(t)$ hold for Lipschitz condition. That is

$$\|\Gamma u_1 - \Gamma u_2\|_\infty \leq M \|u_1 - u_2\|_\infty \qquad (1.20)$$

which implies,

$$\begin{aligned}
\Gamma(u_1) &= \frac{1-\alpha}{AB(\alpha)} f_1(u_1,v,w,t) + \frac{\alpha}{AB(\alpha)\Gamma(\alpha)} \int_0^t f_1(u_1,y,\tau)(t-\tau)^{\alpha-1} d\tau, \\
\Gamma(u_2) &= \frac{1-\alpha}{AB(\alpha)} f_1(u_2,v,w,t) + \frac{\alpha}{AB(\alpha)\Gamma(\alpha)} \int_0^t f_1(u_2,v,w,\tau)(t-\tau)^{\alpha-1} d\tau.
\end{aligned} \qquad (1.21)$$

Consequently,

$$\begin{aligned}
\|\Gamma u_1 - \Gamma u_2\|_\infty &= \frac{1-\alpha}{AB(\alpha)} \|f_1(u_1,v,w,t) - f_1(u_2,v,w,t)\|_\infty \\
&\quad + \frac{\alpha}{AB(\alpha)\Gamma(\alpha)} \|f_1(u_1,v,w,t) - f_1(u_2,v,w,t)\|_\infty \int_0^t (t-\tau) d\tau \\
&\leq \|f_1(u_1,v,w,t) - f_1(u_2,v,w,t)\|_\infty \\
&\quad \left(\frac{1-\alpha}{AB(\alpha)} + \frac{\alpha}{AB(\alpha)\Gamma(\alpha)} \cdot \frac{a^\alpha}{\alpha} \right) \\
&\leq \|f_1(u_1,v,w,t) - f_1(u_2,v,w,t)\|_\infty \left(\frac{1-\alpha}{AB(\alpha)} + \frac{a^\alpha}{AB(\alpha)\Gamma(\alpha)} \right)
\end{aligned} \qquad (1.22)$$

So, the local source term f_1 is Lipschitz continuous with respect to u, if

$$\begin{aligned} \|\Gamma u_1 - \Gamma u_2\|_\infty \leq & K\|u_1 - u_2\|_\infty \left\{ \frac{1-\alpha}{AB(\alpha)} + \frac{a^\alpha}{AB(\alpha)\Gamma(\alpha)} \right\} \\ \leq & L\|u_1 - u_2\|_\infty. \end{aligned} \tag{1.23}$$

This procedure is repeated for reaction kinetics f_2 and f_3 with respect to components v and w to obtain the Lipschitz conditions

$$\begin{aligned} \|\Gamma v_1 - \Gamma v_2\|_\infty \leq & L\|v_1 - v_2\|_\infty, \\ \|\Gamma w_1 - \Gamma w_2\|_\infty \leq & L\|w_1 - w_2\|_\infty. \end{aligned} \tag{1.24}$$

Under this condition, the operation is a contraction on a Banach space $\mathcal{B}$ with norm, which shows that Γ has the property that, $\exists$ a unique function, say $\vartheta : \Gamma\vartheta = \vartheta$, which is the unique solution of the three-species dynamical system (1.14).

1.3 Model Equations and Stability Analysis

In this section, we introduce two dynamical examples consisting of three-species systems with different functional response that are still of current and recurring interests. Such models will be examined for local stability analysis.

1.3.1 Fractional Food Chain Dynamics with Holling Type II Functional Response

The dynamics of predator–prey with two species are often common and easily studied in literature [11, 15, 16, 18–21]. Predators are regarded as top group of species that feed on the lower class called prey. But the real ecosystem is not balanced when description is limited to just two-species system. For instance, snakes feed on mouse. The question is, what does mouse depend on for existence? Meaning that other factors must be considered in addition to predator and prey. There is a popular adage in Yoruba language that says, *oká kìí je oká, oun tí ó ń je okà ni oká ńje*, which means a cobra does not eat grains but eats what eats grain. Hence, we are motivated to formulate a three-species system and examine the amazing dynamics that may occur.

Let us consider a food chain that consists of grasses, mice and cobras, all in the same closed habitat. Denote the grass mass by u, number of mice by v and population of cobras by w. We assume that in the absence of class v, grass would persist and exhibit a logistic growth. We also let that per rabbit consumption rate of grass evolves with increasing grass mass, taking the form of Holling type II functional response $R = \rho_1 u/(1 + \beta_1 u)$. The food web discussed here is mathematically represented by a balanced system of ordinary differential equations

$$\begin{aligned}\frac{du}{dt} &= f_1(u,v,w) = u(1-u) - \frac{\rho_1 u}{1+\beta_1 u}v,\\ \frac{dv}{dt} &= f_2(u,v,w) = \frac{\rho_1 u}{1+\beta_1 u}v - \frac{\rho_2 v}{1+\beta_2 v}w - \delta_1 v,\\ \frac{dw}{dt} &= f_3(u,v,w) = \frac{\rho_2 v}{1+\beta_2 v}w - \delta_2 w\end{aligned} \tag{1.25}$$

where u, v and w denote dimensionless species densities, and $\rho_i > 0, \beta_i > 0$ and $\delta_i > 0$ for $i = 1, 2$ are all positive parameters.

In recent developments, it has been shown that modelling of real-world phenomena with non-integer order derivative is more accurate and reliable when compared with the classical or integer-order case. Hence, the standard time derivative in (1.25) is replaced with fractional-order operator in the form

$$\begin{aligned}\mathcal{D}_t^\alpha u(t) &= f_1(u,v,w) = u(1-u) - \frac{\rho_1 u}{1+\beta_1 u}v,\\ \mathcal{D}_t^\alpha v(t) &= f_2(u,v,w) = \frac{\rho_1 u}{1+\beta_1 u}v - \frac{\rho_2 v}{1+\beta_2 v}w - \delta_1 v,\\ \mathcal{D}_t^\alpha w(t) &= f_3(u,v,w) = \frac{\rho_2 v}{1+\beta_2 v}w - \delta_2 w\end{aligned} \tag{1.26}$$

where $\mathcal{D}_t^\alpha$ is the fractional derivative operator of order α in the sense of the Atangana–Baleanu operators which is expected to satisfy $0 < \alpha \leq 1$.

To examine the steady states of dynamics (1.26), we let

$$\mathcal{D}_t^\alpha u(t) = 0, \quad \mathcal{D}_t^\alpha v(t) = 0, \quad \mathcal{D}_t^\alpha v(t) = 0.$$

Obviously, the system has five equilibrium points. We are only interested in the biologically meaningful interior non-trivial state denoted as $E = (u^*, v^*, w^*)$, where

$$\begin{aligned}u^* &= \frac{\beta_1 - 1}{2b} + \frac{\sqrt{(\beta_1+1)^2 - \frac{4\rho_1\beta_1\delta_2}{\rho_2 - \beta_2\delta_2}}}{2\beta_1}, \quad v^* = \frac{\delta_2}{\rho_2 - \beta_2\delta_2},\\ w^* &= \frac{(\rho_1 - \beta_1\delta_1)u^* - \delta_1}{(\rho_2 - \beta_2\delta_2)(1+\beta_1 u^*)}.\end{aligned}$$

At point E, the community matrix is given as

$$B = \begin{pmatrix} 1 - \frac{\rho_1 v}{(1+\beta_1 u)^2} - 2u & -\frac{\rho_1 u}{(1+\beta_1 u)} & 0\\ \frac{\rho_1 v}{(1+\beta_1 u)^2} & -\frac{\rho_1 v}{(1+\beta_1 u)} - \frac{\rho_2 w}{(1+\beta_2 v)^2} - \delta_1 & -\frac{\rho_2 v}{(1+\beta_1 v)}\\ 0 & \frac{\rho_2 w}{(1+\beta_2 v)^2} & \frac{\rho_2 v}{(1+\beta_1 v)} - \delta_2 \end{pmatrix} \tag{1.27}$$

The eigenvalues corresponding to the interior point E which show the existence of the three species are given by

$$\lambda^3 + \mu_1\lambda^2 + \mu_2\lambda + \mu_3 = 0, \tag{1.28}$$

where

$$\mu_1 = 2u^* + \rho_2\left(\frac{w^*}{(1+\beta_2 v^*)^2} - \frac{v^*}{(1+\beta_2 v^*)}\right)$$
$$+ \rho_1\left(\frac{v^*}{(1+\beta_1 u^*)^2} - \frac{u^*}{(1+\beta_2 u^*)}\right) + \delta_2 + \delta_1 - 1,$$
$$\mu_2 = \rho_1\rho_2\left(\frac{u^*v^*}{(1+\beta_1 u^*)(1+\beta_2 v^*)} - \frac{v^{*2}}{(1+\beta_1 u^*)^2(1+\beta_2 v^*)}\right.$$
$$\left.+\frac{v^*w^*}{(1+\beta_1 u^*)^2(1+\beta_2 v^*)^2}\right)$$
$$+ \rho_2\left(\frac{(2u^*-1+\delta_2)w^*}{(1+\beta_2 v^*)^2} + \frac{(1-2u^*-1+\delta_1)v^*}{(1+\beta_2 v^*)}\right)$$
$$+ \rho_1\left(\frac{(\delta_1+\delta_2)v^*}{(1+\beta_1 v^*)^2} + \frac{(1-2^*-\delta_2)u^*}{(1+\beta_1 u^*)}\right)$$
$$+ (2u^*-1)(\delta_1+\delta_2) + \delta_1\delta_2,$$
$$\mu_3 = \rho_2\left(\frac{(2u^*-1)\delta_2 w^*}{(1+\beta_2 v^*)^2} + \frac{(-2u^*+1)\delta_1 v^*}{(1+\beta_2 v^*)}\right)$$
$$+ \rho_1\left(\frac{\delta_1\delta_2 v^*}{(1+\beta_1 u^*)^2} + \frac{(1-2u^*)\delta_2 u^*}{1+\beta_1 u^*}\right) + 2\delta_1\delta_2 u^*$$
$$\rho_1\rho_2\left(\frac{(2u^*-1)u^*v^*}{(1+\beta_1 u^*)(1+\beta_2 v^*)} - \frac{\delta_1 v^{*2}}{(1+\beta_1 u^*)^2(1+\beta_2 v^*)}\right.$$
$$\left.+\frac{\delta_2 v^*w^*}{(1+\beta_1 u^*)^2(1+\beta_2 v^*)^2}\right) - \delta_1\delta_2.$$

So, following a similar argument [13], we obtained the eigenvalues of the earlier characteristic polynomial as

$$\Lambda = 18\mu_1\mu_2\mu_3 + \mu_1^2\mu_2^2 - 4\mu_3\mu_1^2 - 4\mu_2^3 - 27\mu_3^2.$$

We conclude that if $\Lambda > 0$, the necessary and sufficient condition for the interior point E to be locally and asymptotically stable is that $\mu_1 > 0, \mu_3 > 0$ and $\mu_1\mu_2 - \mu_3 > 0$. But if $\Lambda < 0$, with $\mu_1 > 0, \mu_3 > 0$ and $\mu_1\mu_2 = \mu_3 > 0$, then we say that point E is locally asymptotically stable for values of α in interval $(0, 1)$.

In numerical framework as displayed in the upper row of Figure 1.1, we simulate with initial condition and parameter values

$$\rho_3, \rho_2 = 0.2, \beta_1 = 3, \beta_2 = 2.5, \delta_1 = 0.4, \delta_2 = 0.01,$$
$$(u_0, v_0, w_0) = (0.1, 0.1, 0.1). \tag{1.29}$$

In the experiment, the dynamic behaviour of system (1.26) is observed for different instances of fractional power α as shown in the figure caption. Chaotic patterns are obtained regardless of the value of α chosen in the interval $(0, 1)$.

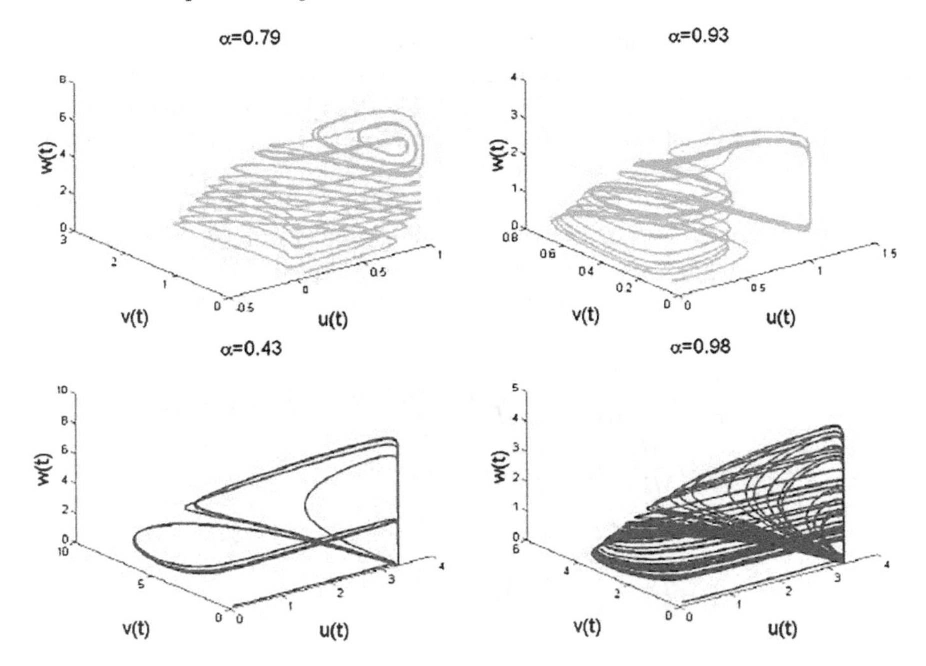

FIGURE 1.1
Chaotic phase portraits for fractional multi-species dynamics (1.26) and (1.32) as shown in rows 1 and 2, respectively, for different α at $t = 1{,}000$. Initial conditions and parameter values are given in (1.29) and (1.38), respectively.

Apart from the distributions reported here, other chaotic and spatiotemporal phenomena are obtainable subject to the choice of initial condition and parameters.

1.3.2 Multi-Species Ecosystem with a Beddington–DeAngelis Functional Response

The dynamics of multi-species ecosystem with a Beddington–DeAgelis functional response whose prey, intermediate-predator and top-predator densities are represented by U, V and W, is governed by the system of differential equations

$$\begin{aligned} \frac{dU}{dt} &= \frac{K(M_1 - U)U}{M_2 - U} - P_1(U,V)W, \\ \frac{dV}{dt} &= A_1P_1(U,V)W - P_2(V,W)W - D_1V, \\ \frac{dW}{dt} &= A_2P_2(V,W)W - D_2W \end{aligned} \tag{1.30}$$

with $P_i(X,Y) = E_iX/(F_iY + G_iX + H_i), i = 1,2$. The Beddington–DeAngelis functional response type is denoted by functions $P_1(U,V)$ and

$P_2(V, W)$. The saturation parameters for these responses are given by E_i, F_i and G_i for $i = 1, 2$. The maximum rate at which predator consumes prey is given by ratio E_i/H_i, while H_i/F_i and $H_i/G_i (i = 1, 2)$ are constants of half saturation. The constants $K, M_i, E_i, F_i, G_i, H_i$ and $D_i (i = 1, 2)$ are assumed positive parameters. M_2 stands for limiting resources, $D_i (i = 1, 2)$ are the death rates of predator class V and W, respectively. The intrinsic growth of prey species is denoted by K, the rate of conversion of prey to either the intermediate predator or top predator is denoted by A_1 and A_2, respectively.

To reduce the number of parameters in system (1.30), we re-scale the model variables as

$$
\begin{aligned}
u = \frac{U}{H_1}, v = \frac{V}{H_1}, w = \frac{W}{H}, t = D_1 T, k = \frac{K}{D_1}, m_1 = \frac{M_1}{H_1}, \\
m_2 = \frac{M_2}{H_1}, \phi_1 = \frac{E_1}{D_1}, \phi_2 = F_1, \\
\phi_3 = G_1, \varphi_1 = \frac{E_1 A_1}{D_1}, \varphi_2 = \frac{E_2}{D_1}, \varphi_3 = F_2, \psi_1 = G_2, \\
\psi_2 = \frac{H_2}{H_1}, \psi_3 = \frac{E_2 A_2}{D_2}, \delta = \frac{D_2}{D_1}
\end{aligned}
$$

to obtain a dimensionless model

$$
\begin{aligned}
\frac{du}{dt} &= f_1(u, v, w) = \frac{k(m_1 - u)u}{m_2 - u} - \frac{\phi_1 uv}{\phi_2 v + \phi_3 u + 1}, \ u(0) > 0, \\
\frac{dv}{dt} &= f_2(u, v, w) = \frac{\varphi_1 uv}{\phi_2 v + \phi_3 u + 1} - \frac{\varphi_2 vw}{\varphi_3 w + \psi_1 v + \psi_2} - v, \ v(0) > 0, \\
\frac{dw}{dt} &= f_3(u, v, w) = \frac{\psi_3 vw}{\varphi_3 w + \phi_3 v + \psi_2} - \delta w, \ w(0) > 0
\end{aligned}
\tag{1.31}
$$

where $f_i(u, v, w)$ are functions representing the local kinetics, and u, v and w are functions of time.

Our interest here is to model the tritrophic system (1.31) with non-integer order derivative in the form

$$
\begin{aligned}
\mathcal{D}_t^\alpha u &= f_1(u, v, w) = \frac{k(m_1 - u)u}{m_2 - u} - \frac{\phi_1 uv}{\phi_2 v + \phi_3 u + 1}, \\
\mathcal{D}_t^\alpha v &= f_2(u, v, w) = \frac{\varphi_1 uv}{\phi_2 v + \phi_3 u + 1} - \frac{\varphi_2 vw}{\varphi_3 w + \psi_1 v + \psi_2} - v, \\
\mathcal{D}_t^\alpha w &= f_3(u, v, w) = \frac{\psi_3 vw}{\varphi_3 w + \phi_3 v + \psi_2} - \delta w,
\end{aligned}
\tag{1.32}
$$

where $\mathcal{D}_t^\alpha$ denotes fractional derivative of order $\alpha \in (0, 1]$. It should be noted that we recover the standard derivative whenever α tends to unity.

With $\mathcal{D}_t^\alpha u = 0, \mathcal{D}_t^\alpha v = 0$ and $\mathcal{D}_t^\alpha w = 0$, it is not difficult to see that system (1.32) has four equilibrium points: they are $E^0 = (0, 0, 0)$ which is the total washout state and $E^1 = (m_1, 0, 0)$ which shows the existence of the prey only.

The point $\bar{E} = (\bar{u}, \bar{v}, 0)$, where $\bar{v} = \frac{\bar{u}(\varphi_1 - \phi_3) - 1}{\phi_2}$ and $\bar{u} = \frac{\gamma_2 \pm \sqrt{\gamma_4}}{\gamma_1}$ subject to conditions $\gamma_1 > 0, \gamma_4 > 0$ or $\gamma_1 < 0, \gamma_4 > 0$, respectively, shows that both prey and intermediate predator can survive in the absence of top predator. The non-trivial point $E^* = (u^*, v^*, w^*)$ is referred to as the balance coexistence state, where

$$v^* = \frac{\delta(\varphi_3 w^* + \psi_2)}{\psi_3 - \psi_1 \delta}, w^* = \frac{k\varphi_1(\psi_3^2 - \psi_1\psi_3\delta)(m_1 - u^*)u^*}{\xi_1(m_1 - u^*)} - \frac{\xi_2}{\xi_1}.$$

and

$$\frac{\xi_3 - \sqrt{\xi_4}}{2k\varphi_1\psi_3(\psi_3 - \psi_1\delta)} < u^* < \frac{\xi_3 + \sqrt{\xi_4}}{2k\varphi_1\psi_3(\psi_3 - \psi_1\delta)},$$

where

$$\xi_1 = \phi_1\varphi_3\psi_3\delta + \phi_1\varphi_2\delta(\psi_3 - \psi_1\delta), \quad \xi_2 = \phi_1\psi_2\psi_3\delta,$$
$$\xi_3 = \xi_2 + k\varphi_1\psi_3(\psi_3 - \psi_1\delta)m_1, \quad \xi_4 = \xi_3^2 - 4k\varphi_1\psi_3(\psi_3 - \psi_1\delta)\xi_2 m_2.$$

For the stability analysis, we will only consider the last two equilibrium states that are biologically meaningful, since most ecosystems usually consist of either the prey and predator or the addition of a top predator. The local stability of system (1.32) at point $\bar{E} = (\bar{u}, \bar{v}, 0)$, the resulting equation reduces to system

$$\begin{aligned} \mathcal{D}_t^\alpha u &= \frac{k(m_1 - u)u}{m_2 - u} - \frac{\phi_1 uv}{\phi_2 v + \phi_3 u + 1}, \\ \mathcal{D}_t^\alpha v &= \frac{\varphi_1 uv}{\phi_2 v + \phi_3 u + 1} - v. \end{aligned} \tag{1.33}$$

The community matrix of system (1.33) is defined by

$$A = \left(\begin{array}{cc} \underbrace{u^*\left[\frac{k(m_1 - m_2)}{(m_2 - u^*)^2} + \frac{\phi_1\phi_3 v^*}{(\phi_2 v^* + \phi_3 u^* + 1)^2}\right]}_{s_{11}} & \underbrace{\frac{-\phi_1 u^*(\phi_3 + 1)}{(\phi_2 v^* + \phi_3 u^* + 1)^2}}_{s_{12}} \\ \underbrace{\frac{\varphi_1 v^*(\phi_2 + 1)}{(\phi_2 v^* + \phi_3 u^* + 1)^2}}_{s_{21}} & \underbrace{\frac{-\phi_1\varphi_1 u^* v^*}{(\phi_2 v^* + \phi_3 u^* + 1)^2}}_{s_{22}} \end{array} \right)_{(u^*, v^*)}. \tag{1.34}$$

The characteristic equation of matrix A is

$$\lambda^2 - (s_{11} + s_{22})\lambda + (s_{11}s_{22} - s_{12}s_{21}).$$

By adopting the Routh–Hurwitz criterion, it is easily verified that the point $\bar{E}$ of a subsystem (1.33) is locally asymptotically stable provided, both $-(s_{11} + s_{22}) > 0$ and $(s_{11}s_{22} - s_{12}s_{21}) > 0$.

The community matrix of system (1.32) at interior point $E^* = (u^*, v^*, w^*)$ is defined by

$$C = \begin{pmatrix} c_{11} & c_{12} & c_{13} \\ c_{21} & c_{22} & c_{23} \\ c_{31} & c_{32} & c_{33} \end{pmatrix}_{(u^*,v^*,w^*)} \tag{1.35}$$

where

$$\begin{aligned} c_{11} &= u^* \left[\frac{k(m_1 - m_2)}{(m_2 - u^*)^2} + \frac{\phi_1 \phi_3 v^*}{(\phi_2 v^* + \phi_3 u^* + 1)^3} \right], \\ c_{12} &= \frac{-\phi_1 u^* (\phi_3 + 1)}{(\phi_2 v^* + \phi_3 u^* + 1)^2} < 0, \\ c_{21} &= \frac{\varphi_1 v^* (\phi_3 + 1)}{(\phi_2 v^* + \phi_3 u^* + 1)^2} > 0, \\ c_{22} &= v^* \left[\frac{-\phi_2 \varphi_1 u^*}{(\phi_2 v^* + \phi_3 u^* + 1)^2} + \frac{\varphi_2 \psi_1 w^*}{(\varphi_3 w^* + \psi_1 v^* + \psi_2)^2} \right] \\ c_{23} &= \frac{-\varphi_2 v^* (\psi_1 v^* + \psi_2)}{(\varphi_3 w^* + \psi_1 v^* + \psi_2)^2} < 0; c_{32} = \frac{-\psi_3 w^* (\varphi_3 w^* + \psi_2)}{(\varphi_2 w^* + \psi_1 v^* + \psi_2)^2} > 0 \\ c_{33} &= -\frac{-\varphi_3 \psi_3 v^* w^*}{(\varphi_3 w^* + \psi_1 v^* + \psi_2)^2} < 0, c_{13} = 0, \ c_{31} = 0. \end{aligned} \tag{1.36}$$

The characteristic equation at steady state $C(u^*, v^*, w^*)$ is

$$\begin{aligned} &\lambda^3 - (c_{11} + c_{22} + c_{33})\lambda^2 + (c_{11}c_{22} - c_{12}c_{21} + c_{22}c_{33} - c_{23}c_{32} \\ &\quad + c_{11}c_{33})\lambda + (c_{11}c_{23}c_{32} - c_{11}c_{22}c_{33} + c_{12}c_{21}c_{33}). \end{aligned} \tag{1.37}$$

So, if conditions $-(c_{11}+c_{22}+c_{33})\lambda^2 > 0$, $(c_{11}c_{23}c_{32}-c_{11}c_{22}c_{33}+c_{12}c_{21}c_{33}) > 0$ and $(-(c_{11}+c_{22}+c_{33}))(c_{11}c_{22}-c_{12}c_{21}+c_{22}c_{33}-c_{23}c_{32}+c_{11}c_{33}) > (c_{11}c_{23}c_{32} - c_{11}c_{22}c_{33} + c_{12}c_{21}c_{33})$ hold, then the non-trivial point $E^* = (u^*, v^*, w^*)$ is locally asymptotically stable.

In the simulation experiments, the dynamical behaviour of a fractional three-species system (1.32) is examined with the following parameters and initial condition

$$\begin{aligned} &k = 3, \phi_1 = 4, \phi_2 = 0.2, \phi_3 = 0.8, \varphi_1 = 3, \varphi_2 = 2, \varphi_3 = 0.15, \psi_1 = 0.8, \\ &\psi_2 = 0.5, \psi_3 = 1.5, \delta = 0.2, m_1 = 3.33, m_2 = 4, (u_0, v_0, w_0) = (0.1, 0.1, 0.1). \end{aligned} \tag{1.38}$$

The results of this experiment are given in the second row of Figure 1.1, which correspond to species chaotic attractor result. It was observed in the simulation that the number of periodic oscillations increases with increase in the value of α.

1.4 Numerical Experiment for Fractional Reaction-Diffusion Ecosystem

We have examined the behaviour of non-spatial in the previous section. In this segment, the dynamical behaviour of three-species fractional reaction-diffusion ecosystem will be examined numerically. The numerical simulations are experimented with MATLAB® software package. Investigation is based mainly on the effect of fractional power $\alpha \in (0, 1)$.

Let us consider the fractional reaction-diffusion ecological system

$$\begin{aligned}
\mathcal{D}_t^\alpha u &= \frac{\partial^2 u}{\partial x^2} + \frac{k(m_1 - u)u}{m_2 - u} - \frac{\phi_1 uv}{\phi_2 v + \phi_3 u + 1}, \\
\mathcal{D}_t^\alpha v &= \frac{\partial^2 v}{\partial x^2} + \frac{\varphi_1 uv}{\phi_2 v + \phi_3 u + 1} - \frac{\varphi_2 vw}{\varphi_3 w + \psi_1 v + \psi_2} - v, \\
\mathcal{D}_t^\alpha w &= \frac{\partial^2 w}{\partial x^2} + \frac{\psi_3 vw}{\varphi_3 w + \phi_3 v + \psi_2} - \delta w,
\end{aligned} \tag{1.39}$$

subject to homogeneous (zero-flux) Neumann boundary conditions mount at the extreme of domain size $[0, L], L > 0$ and the initial condition

$$\begin{aligned}
u(x, 0) &= u^*(ones(N, 1)), \\
v(x, 0) &= v^*(ones(N, 1)), \\
w(x, 0) &= w^*(ones(N, 1)),
\end{aligned} \tag{1.40}$$

specifically chosen to induce a non-trivial solution. The states (u^*, v^*, w^*) in (1.40) correspond to $(0.07, 0.01, 0.02)$, and N is the number of discretization points. The second-order partial derivative is approximated with either the fourth-order central difference scheme

$$\frac{\partial^2 \nu}{\partial x^2} \approx \frac{-\nu_{i-2,j} + 16\nu_{i-1,j} - 30\nu_{i,j} + 16\nu_{i+1,j} - \nu_{i+2,j}}{12h^2}, \tag{1.41}$$

or the Fourier spectral method [20].

Let us consider the spatial discretization of function z on grid $x_1, x_2, \ldots, z_N$ with $z_j = z(x_j)$. Assuming z is periodic, we introduce the discrete Fourier transform (DFT) technique $\hat{z}_k = \hat{z}(k)$ defined by

$$\hat{z}_k = \sum_{j=i}^{N} e^{-ikx_j} z_j, \quad k = -\frac{N}{2} + 1, \ldots \frac{N}{2}, \tag{1.42}$$

where i the complex unit and k denotes the Fourier wave numbers. Similarly, we define the inverse discrete Fourier transform (IDFT) as

$$z_j = \frac{1}{2\pi} \sum_{k=-N/2+1}^{N} e^{ikx_j} \bar{z}_j, \quad j = 1, 2, \ldots, N. \tag{1.43}$$

If the approximation of z at x_j is represented by ω_j, then

$$\omega_j = \frac{dz}{dx}(x_j), \quad j = 1, 2, \ldots, N. \tag{1.44}$$

Given z_j, $j = 1, 2, \ldots, N$, by applying (1.42), one computes the DFT $\hat{z}_k$, $k = -N/2+1, \ldots, N/2$. If z_j is defined to be $ik\hat{z}_k$, then for $k = -N/2+1, \ldots, N/2$, we use (1.43) to compute z_j, $j = 1, 2, \ldots, N$ from $\hat{z}_k$, for $k = -N/2+1, \ldots, N/2$.

By using the earlier procedure, the fractional reaction-diffusion problem fully transforms a system of ordinary differential equations

$${}_0^{ABC}\mathcal{D}^{\alpha}\hat{\nu}_k = -\delta k^2 \hat{\nu}_k + \widehat{\mathcal{F}(\nu)}, \quad k = -N/2+1, \ldots, N/2 \tag{1.45}$$

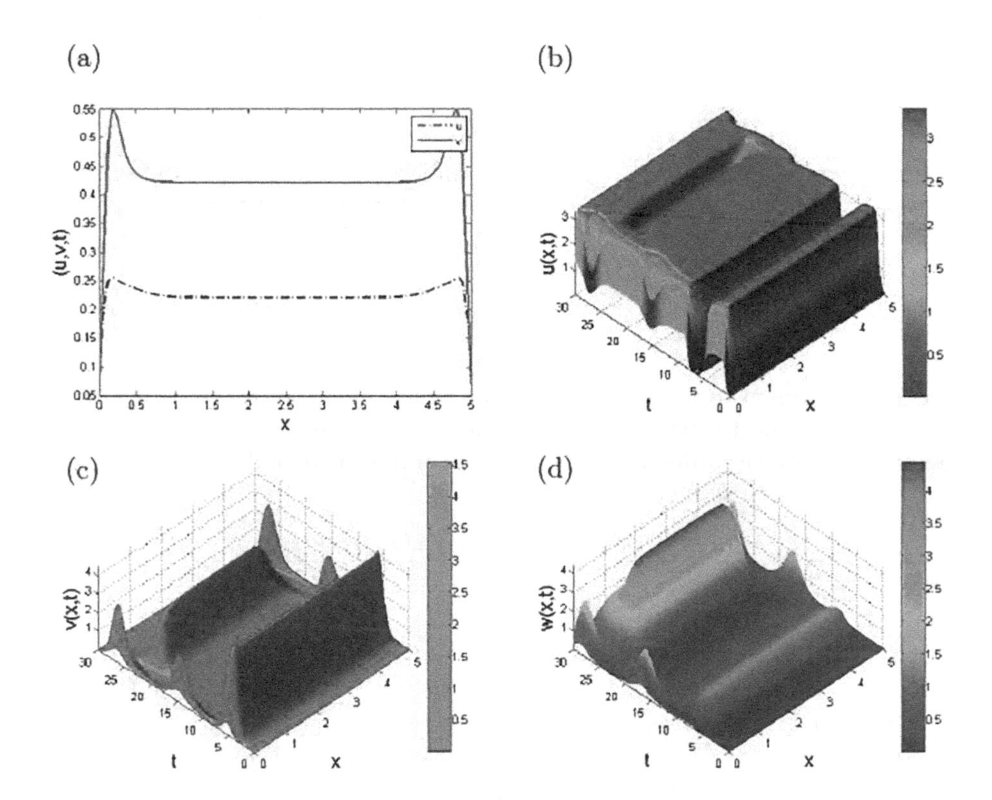

FIGURE 1.2
Chaotic distribution of fractional diffusive ecosystem (1.39) showing effect of $k = 2.5$ for $\alpha = 0.84$ and parameters $\phi_1 = 4, \phi_2 = 0.2, \phi_3 = 0.8$, $\varphi_1 = 3, \varphi_2 = 2, \varphi_3 = 0.15, \psi_1 = 0.8, \psi_2 = 0.5, \psi_3 = 1.5, \delta = 0.2, m_1 = 3.33$, $m_2 = 4$. Plots (a) corresponds to $\bar{E} = (\bar{u}, \bar{v}, 0)$ showing the existence of prey and intermediate predator only, while surface plots (b–d) represent the interior point $E^* = (u^*, v^*, w^*)$.

where $\nu = (\nu_1, \nu_2, \ldots, \nu_n)$ is calculated by (1.43) with $\hat{z}_k$, $k = -N/2 + 1, \ldots, N/2$. In this case, the linear diffusive term in (1.39) has become a diagonal matrix with the issue of stiffness being removed. The main advantage of applying the Fourier spectral method is that any explicit time integrator can be used once the partial differential equations (PDEs) have been converted into ordinary differential equations (ODEs).

In the experiment, we utilize the time step $\Delta t = 0.5$ and spatial step size $\Delta x = h = 0.25$ with parameters $\phi_1 = 4, \phi_2 = 0.2, \phi_3 = 0.8, \varphi_1 = 3,$ $\varphi_2 = 2, \varphi_3 = 0.15, \psi_1 = 0.8, \psi_2 = 0.5, \psi_3 = 1.5, \delta = 0.2, m_1 = 2.85,$ $m_2 = 4, k = 5$ to obtain the numerical results as displayed in Figures 1.2 and 1.3.

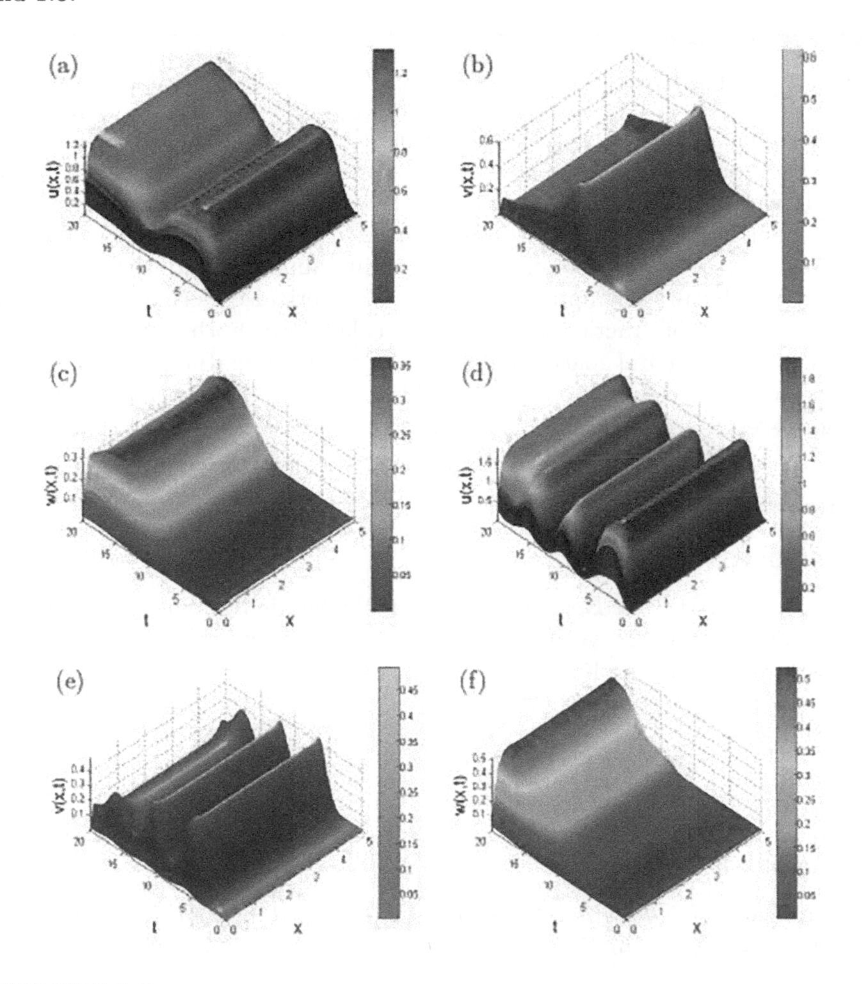

FIGURE 1.3
Chaotic distribution of fractional diffusive ecosystem (1.39) obtained at $m_1 = 1.28$ for different $\alpha = (0.81, 0.89)$ for plots (a–c) and (d–f), respectively. Other parameters are given in Figure 1.2.

Chaotic evolution for fractional reaction-diffusion ecosystem (1.39) at equilibrium point $\bar{E} = (\bar{u}, \bar{v}, 0)$, which corresponds to the existence the prey and intermediate predator species obtained for $\alpha = 0.84$ and parameters $\phi_1 = 4, \phi_2 = 0.2, \phi_3 = 0.8, \varphi_1 = 3, \varphi_2 = 2, \varphi_3 = 0.15, \psi_1 = 0.8, \psi_2 = 0.5, \psi_3 = 1.5, \delta = 0.2, m_2 = 4, k = 5$, is shown in the first and second rows of Figure 1.2. Numerical results in rows 3–5 correspond to the dynamics of the interior equilibrium point $E^* = (u^*, v^*, w^*)$. Parameters at this point are given in the figure caption. Spatiotemporal evolution of the interior point for $\alpha = 0.81$ and $\alpha = 0.89$ at $m_1 = 1.28$ is presented in Figure 1.3. In both cases, the species oscillate in space regardless of α.

1.5 Conclusion

Two dynamics of three-species models consisting of prey, intermediate predator and top predator are considered in this work. In the model, the classical time derivatives are replaced with the new Atangana–Baleanu fractional derivative in the sense of Caputo. To guarantee the correct choice of parameters, such systems are examined for stability analysis and they are confirmed to be locally asymptotically stable. Numerical experiments obtained for some values of fractional order α revealed that both systems exhibit chaotic phenomena and spatiotemporal oscillations for spatial and non-spatial cases. The methodology reported in this work can be extended to solve fractional-reaction-diffusion problems in high dimensions.

References

[1] Aghababa, M. P. & Borjkhani, M. (2014). Chaotic fractional-order model for muscular blood vessel and its control via fractional control scheme. *Complexity*, 20, 37–46.

[2] Allen, L. J. S., *An Introduction to Mathematical Biology*, Pearson Education, New Jersey, 2007.

[3] Atangana, A. & Baleanu, D. (2016). New fractional derivatives with non-local and non-singular kernel: Theory and application to heat transfer model. *Thermal Science*, 20, 763–769.

[4] Atangana, A., *Fractional Operators with Constant and Variable Order with Application to Geo-hydrology*, Academic Press, New York, 2017.

[5] Atangana, A. (2018). Non validity of index law in fractional calculus: A fractional differential operator with Markovian and non-Markovian properties. *Physica A: Statistical Mechanics and its Applications*, 505, 688–706.

[6] Atangana, A. & Owolabi, K. M. (2018). New numerical approach for fractional differential equations. *Mathematical Modelling of Natural Phenomena*, 13(3), 1–21.

[7] Benson, D. A., Meerschaert, M. M. & Revielle, J. (2013). Fractional calculus in hydrologic modeling: A numerical perspective. *Advances in Water Resources*, 51, 479–497.

[8] Caputo, M. & Fabrizio, M. (2015). A new definition of fractional derivative without singular kernel. *Progress in Fractional Differentiation and Applications*, 1, 73–85.

[9] El-Misiery, A. E. M. & Ahmed, E. (2006). On a fractional model for earthquakes. *Applied Mathematics and Computation*, 178, 207–211.

[10] Garvie, M. R. & Trenchea, C. (2009). Spatiotemporal dynamics of two generic predator-prey models. *Journal of Biological Dynamics*, 4, 1–12.

[11] Hastings, A. & Powell, T. (1991). Chaos in a three-species food chain. *Ecology*, 72, 896–903.

[12] Lotka, A. J. *Elements of Physical Biology*, Williams and Wilkins, Baltimore, 1925.

[13] Matouk, A. E., Elsadany, A. A., Ahmed, E. & Agiza, H. N. (2015). Dynamical behavior of fractional-order Hastings-Powell food chain model and its discretization. *Communications in Nonlinear Science and Numerical Simulation*, 27, 153–167.

[14] Meral, F. C., Royston, T. J. & Magin, R. (2010). Fractional calculus in viscoelasticity: An experimental study. *Communications in Nonlinear Science and Numerical simulation*, 15, 939–945.

[15] Murray, J. D., *Mathematical Biology I: An Introduction*, Springer-Verlag, New York, 2002.

[16] Murray, J. D., *Mathematical Biology II: Spatial Models and Biomedical Applications*, Springer-Verlag, Berlin, 2003.

[17] Owolabi, K. M. & Patidar, K. C. (2014). Numerical solution of singular patterns in one-dimensional Gray-Scott-like models. *International Journal of Nonlinear Science and Numerical Simulations*, 15, 437–462.

[18] Owolabi, K. M. & Patidar, K. C. (2016) Numerical simulations of multicomponent ecological models with adaptive methods. *Theoretical Biology and Medical Modelling*, 13, 1.

[19] Owolabi, K. M. (2016). Mathematical study of two-variable systems with adaptive numerical methods. *Numerical Analysis and Applications*, 9, 218–230.

[20] Owolabi, K. M. (2017). Mathematical study of multispecies dynamics modeling predator-prey spatial interactions. *Journal of Numerical Mathematics*, 25, 1–16.

[21] Owolabi, K. M. & Pindza, E. (2018). Mathematical and computational studies of fractional reaction-diffusion system modelling predator-prey interactions. *Journal of Numerical Mathematics*, 26, 97–110.

[22] Owolabi, K. M. (2019). Computational study of noninteger order system of predation. *Chaos*, 29, 013120.

[23] Owolabi, K. M. & Hammouch, Z. (2019). Mathematical modeling and analysis of two-variable system with noninteger-order derivative. *Chaos*, 29, 013145.

[24] Owolabi, K. M. & Atangana, A. (2019). On the formulation of Adams-Bashforth scheme with Atangana-Baleanu-Caputo fractional derivative to model chaotic problems. *Chaos*, 29, 023111.

[25] Petrovskii, S., Li, B. & Malchow, H. (2003). Quantification of the spatial aspect of chaotic dynamics in biological and chemical systems. *Bulletin of Mathematical Biology*, 65, 425–446.

[26] Pindza, E. & Owolabi, K. M. (2016). Fourier spectral method for higher order space fractional reaction-diffusion equations. *Communications in Nonlinear Science and Numerical simulation*, 40, 112–128.

[27] Podlubny, I., *Fractional Differential Equations*, Academic Press, San Diego, 1999.

[28] Sun, G., Zhang, G., Jin, Z. & Li, L. (2009). Predator cannibalisms can give rise to regular spatial patterns in a predator-prey system. *Nonlinear Dynamics*, 58, 75–84.

[29] Volterra, V. (1926). Variazioni e fluttuazioni del numero di individui in specie animali conviventi. *Mem Acad Lincei*, 2, 31–113.

2

Solutions for Fractional Diffusion Equations with Reactive Boundary Conditions

Ervin K. Lenzi

Universidade Estadual de Ponta Grossa

Marcelo K. Lenzi

Universidade Federal do Paraná

CONTENTS

2.1 Introduction

In the real life, many processes of practical interest in engineering [1,2], biological systems [3], detection of disease, and physics [4–6] have the surface as a fundamental structure. They may involve molecule diffusion and reaction processes with creation or annihilation of particles. For example, industrial and biochemical reactions can have the reaction rate or reagent sorption limited by the mass transfer between the fluid phase and the catalyst surface [7,8]. These processes may also be limited by the desorption rate and transfer of reaction product to the fluid phase, which occurs in the opposite direction to the reagent molecules that are diffusing from the interior of the fluid to the particle. In these contexts, the heterogeneous reaction systems play an important role in an industrial scenario. They occur on the interface of different materials with reactants at different catalytic systems, e.g., gas–solid or liquid–solid. In biological systems [9], the surfaces, i.e., membranes, are responsible for the selectivity of particles by means of sorption and desorption processes and, consequently, the particles transfer from one region to the other [10–13]. This dynamic cycle is crucial for maintaining life, including simple diffusion, facilitated diffusion, osmosis and active transport [9]. Other contexts can also be found in physics such as the electrical response of water [14] or liquid crystals [15], where the effects on the surface of the interface between electrode and fluid play an important role essentially in the low frequency limit, where the impedance is characterized by the behaviour $\mathcal{Z} \sim 1/(i\omega)^\gamma$ with $0 < \gamma < 1$. This behaviour is the same of the ones exhibited by the constant phase elements.

These scenarios are usually described in terms of differential equations, and in general, are connected to the Markovian processes [16] or Debye relaxations. Differential equations play an important role in describing both time- and space-dependent variables, usually resulting from mass, energy and momentum conservation laws. However, different aspects such as the morphology of the surfaces, i.e., fractal characteristics, interaction between the particles, and memory effects, have evidenced the limitations of the usual approach in describing a large variety of situations and, consequently, leading to non-Fickian diffusion scenarios [3, 17, 18], which are characterized by a non-linear dependence on the mean square displacement. These features have challenged several researchers to investigate alternative approaches in order to provide a suitable analysis of these phenomena. For instance, continuous time random walk [19], generalized Langevin equations and fractional diffusion equations (see Refs. [20–29]) have been successfully analyzed and applied in several situations such as kinetic of surfaces [18], molecular diffusion [30], in membrane cells [31, 32], subdiffusion in thin membranes [33] and electrical response [34]. In these situations, one of the main issues is the non-linear time dependence exhibited by the mean-square displacement, which, in general, is characterized by $(\Delta r)^2(t) \sim t^\alpha$, the superdiffusion being usually related, for example, to active transport [35], while the subdiffusive behaviour may be related to the molecular crowding [36] and fractal structure [26, 37]. In Ref. [38], fractional diffusion equations are used to study the desorption process of methane in coal. In addition, these situations require a suitable choice of the boundary conditions [39] to allow an actual description of these phenomena in connection with the specific processes undergone by particles in the vicinity and on the system surface. From a classical point of view, Dirichelet, Neumann and Robin boundary conditions are frequently used to describe actual boundary effects.

Here, we investigate a sorption phenomenon followed by a reaction process that may occur on a surface in contact with a system composed by two different kinds of particles, 1 and 2. We consider that the particles, of the system, in the bulk are governed by fractional diffusion equations. In particular, these generalized diffusion equations may be related to the fractional diffusion equations depending on the conditions considered to describe the bulk dynamics. The processes on the surface are also assumed to be related to linear kinetic equations with memory effects, which may be connected to a non-usual (or non-Debye) relaxation. In this scenario, we may obtain the behaviour of the quantities on the surface where the processes are present in bulk. This development is discussed in Section 2.2. A summary of the results and our conclusions are presented in Section 2.3.

2.2 The Problem: Diffusion and Kinetics

Let us start our discussion by defining the region where the particles are diffusing. It is a semi-infinity medium with a surface located at the origin, where the particles 1 and 2 (species or substances) may be adsorbed, desorbed and/or absorbed. The particles absorbed, i.e., which are removed from the system, by the surface may also react and promote the formation of each other. For simplicity, we consider the diffusion in one dimension, e.g., at the x-direction, with the particles in the bulk governed by the fractional diffusion equations

$$\frac{\partial}{\partial t}\rho_1(x,t) = \mathcal{K}_1\ {}_0\mathcal{F}_{\alpha_1}\left[\frac{\partial^2}{\partial x^2}\rho_1(x,t)\right], \tag{2.1}$$

and

$$\frac{\partial}{\partial t}\rho_2(x,t) = \mathcal{K}_2\ {}_0\mathcal{F}_{\alpha_2}\left[\frac{\partial^2}{\partial x^2}\rho_2(x,t)\right], \tag{2.2}$$

for $0 < x < \infty$. In Eqs. (2.1) and (2.2), $\mathcal{K}_1$ and $\mathcal{K}_2$ are the generalized diffusion coefficients related to the particles 1 and 2, respectively. The quantities ρ_1 and ρ_2 represent the density of each particles present in the bulk, and $\mathcal{F}_{\alpha_{1(2)}}\{\cdots\}$ in Eqs. (2.1) and (2.2) is the fractional operator,

$${}_0\mathcal{F}_{\alpha_{1(2)}}\{\rho_{1(2)}(x,t)\} = \frac{\partial}{\partial t}\int_0^t dt'\Phi_{1(2)}(t-t')\rho_{1(2)}(x,t'), \tag{2.3}$$

where the $\Phi_{1(2)}(t)$ defines the fractional operator to be considered or the memory effect related to the system under analysis. For example, the case characterized by the power law

$$\Phi_{1(2)}(t) = t^{\alpha_{1(2)}-1}/\Gamma\left(\alpha_{1(2)}\right) \tag{2.4}$$

corresponds to the well-known Riemann–Liouville fractional operator [40] for $0 < \alpha_{1(2)} < 1$ and

$$\Phi_{1(2)}(t) = \int_0^1 d\alpha_{1(2)}p(\alpha_{1(2)})\frac{t^{\alpha_{1(2)}-1}}{\Gamma\left(\alpha_{1(2)}\right)} \tag{2.5}$$

leads us to fractional operators of distributed order, where $p(\alpha_{1(2)})$ is a distribution related to the fractional parameter $\alpha_{1(2)}$. Another possible functions for the kernels are, for example, the exponentials [41] or Mittag-Leffler functions [42]. Typical situations with these functions are given by

$$\Phi_{1(2)}(t) = \mathcal{R}(\alpha_{1(2)})\, e^{-\overline{\alpha}_{1(2)}t} \tag{2.6}$$

and

$$\Phi_{1(2)}(t) = \mathcal{R}(\alpha_{1(2)})\, E_{\alpha_{1(2)}}\left(-\overline{\alpha}_{1(2)}t^{\alpha_{1(2)}}\right), \tag{2.7}$$

In Eqs. (2.6) and (2.7), $\mathcal{R}(\alpha_{1(2)})$ is a normalization factor, $\overline{\alpha}_{1(2)} = \alpha_{1(2)}/(1-\alpha_{1(2)})$, and

$$E_{\alpha}(x) = \sum_{n=0}^{\infty} \frac{x^n}{\Gamma(\beta + \alpha n)} \tag{2.8}$$

is the Mittag-Leffler function [40], which is asymptotically governed by a power law, i.e., $E_{\alpha}(x) \sim 1/x$. These choices for $\Phi_{1(2)}(t)$ show that the Riemann–Liouville operator has a singularity at the origin ($t = 0$), while the others are non-singular operators [41–45], and may also manifest different regimes of diffusion as discussed in Ref. [40].

On the surface, we consider that the processes are governed by the following equations:

$$\begin{aligned} \mathcal{K}_1 \frac{\partial}{\partial x}\, {}_0\mathcal{F}_{\alpha_1}\left\{\rho_1(x,t)\right\}\Big|_{x=0} &= \frac{d}{dt}\mathcal{C}_1(t) + \int_0^t k_{11}(t-t')\rho_1(0,t')dt' \\ &\quad - \int_0^t k_{12}(t-t')\rho_2(0,t')dt' \end{aligned} \tag{2.9}$$

and

$$\begin{aligned} \mathcal{K}_2 \frac{\partial}{\partial x}\, {}_0\mathcal{F}_{\alpha_2}\left\{\rho_2(x,t)\right\}\Big|_{x=0} &= \frac{d}{dt}\mathcal{C}_2(t) + \int_0^t k_{22}(t-t')\rho_2(0,t')dt' \\ &\quad - \int_0^t k_{21}(t-t')\rho_1(0,t')dt'. \end{aligned} \tag{2.10}$$

where $k_{11}(t)$, $k_{12}(t)$, $k_{21}(t)$ and $k_{22}(t)$ are related to the rate of particles absorbed and released by the surface. In these equations, $\mathcal{C}_1(t)$ and $\mathcal{C}_2(t)$ represent the density of particles on the surface that are adsorbed or desorbed. For the adsorption and desorption processes on the surface, we assume that they may be modelled by the following equations:

$$\mathcal{C}_1(t) = \mathcal{C}_{0,1}(t) + \int_0^t \kappa_1(t-t')\rho_1(0,t')dt', \tag{2.11}$$

and

$$\mathcal{C}_2(t) = \mathcal{C}_{0,2}(t) + \int_0^t \kappa_2(t-t')\rho_2(0,t')dt'. \tag{2.12}$$

In Eqs. (2.11) and (2.12), $\rho_1(0,t)$ and $\rho_2(0,t)$ are the bulk density just in front of the surface, respectively. The kernel $\kappa_{1(2)}(s)$ is connected to the adsorption–desorption phenomena. For instance, the choice $\kappa_{1(2)}(s) = (\kappa)\, e^{-t/\tau}$ for the kernel may be related to the kinetic equation

$$\tau \frac{d}{dt}\mathcal{C}_{1(2)}(t) = \kappa\tau\rho_{1(2)}(0,t) - \mathcal{C}_{1(2)}(t) \tag{2.13}$$

for a suitable $\mathcal{C}_{0,1(2)}(t)$. Another choices for $\kappa_i(s)$ imply in different kinetic equations for the processes on the surface, such as the case

$$\kappa_{1(2)}(t) = \kappa\left(\frac{t}{\tau}\right)^{\gamma-1} E_{\gamma,\gamma}\left[-\left(\frac{t}{\tau}\right)^{\gamma}\right], \tag{2.14}$$

where

$$E_{\alpha,\beta}(x) = \sum_{n=0}^{\infty} \frac{x^n}{\Gamma(\beta+\alpha n)} \tag{2.15}$$

is the generalized Mittag-Leffler function. The kernel given by Eq. (2.14) is connected to the fractional kinetic equation

$$\tau^{\gamma}\frac{d^{\gamma}}{dt^{\gamma}}\mathcal{C}_{1(2)}(t) = \kappa\tau\rho_{1(2)}(0,t) - \mathcal{C}_{1(2)}(t), \tag{2.16}$$

with the fractional operator defined as follows:

$$\frac{d^{\gamma}}{dt^{\gamma}}\mathcal{C}_{1(2)}(t) = \frac{1}{\Gamma(1-\gamma)}\int_0^t d\bar{t}\frac{\mathcal{C}'_{1(2)}(\bar{t})}{(t-\bar{t})^{\gamma-1}} \tag{2.17}$$

for $0 < \gamma < 1$, where $\mathcal{C}'_{1(2)}(t) \equiv d\mathcal{C}_{1(2)}(t)/dt$ corresponds to the first time derivative, respectively. Thus, Eqs. (2.11) and (2.12) state that the time variation of the surface density of particles on the surface depends on the bulk density of particles just in front of the surface and on the surface density of particles already sorbed. They may be considered extensions of the usual kinetic equations (Langmuir approximation) to situations characterized by non-usual relaxations, i.e., non-Debye relaxations for which a non-exponential behaviour of the densities can be obtained, depending on the choice of the kernels [47–49]. The underlying physical motivation of the time-dependent rate coefficients can be related with the fractal nature, low-dimensionality or macromolecular crowding of the medium, and even with the anomalous molecular diffusion. From a phenomenological point of view, the choice of the kernels of Eqs. (2.11) and (2.12) can be related to surface irregularities [50], which are important in adsorption–desorption, to diffusion, to catalysis processes, and to microscopic parameters representing the van der Waals interaction between the particles and the surfaces. It is also interesting to mention that Eqs. (2.24) and (2.25) are coupled with Eqs. (2.11) and (2.12) in such way that processes occurring for each specie modifies the dynamic of the other side. The searched solutions are also subjected to homogeneous boundary conditions $\partial_x\rho_1(\infty,t) = 0$ and $\partial_x\rho_2(\infty,t) = 0$.

By using the boundary conditions, it is possible to analyze the behaviour of the particles in the bulk, i.e., how the changes of one specie influences the other specie. In particular, after performing some calculations, we have that

$$\frac{d}{dt}\left(\int_0^\infty \rho_1(x,t)dx + \mathcal{C}_1(t)\right) = \int_0^t k_{12}(t-t')\rho_2(0,t')dt' - \int_0^t k_{11}(t-t')\rho_1(0,t')dt', \qquad (2.18)$$

$$\frac{d}{dt}\left(\int_0^\infty \rho_2(x,t)dx + \mathcal{C}_2(t)\right) = \int_0^t k_{21}(t-t')\rho_1(0,t')dt' - \int_0^t k_{22}(t-t')\rho_2(0,t')dt'. \qquad (2.19)$$

where some terms, e.g., $k_{11}(t)$ or $k_{22}(t)$, imply a withdrawal of particles from the bulk, and others, e.g., $k_{12}(t)$ or $k_{21}(t)$, are related to the release of particles from the surface to the bulk. For the particular case $k_{11}(t) = k_{21}(t) = k_1(t)$ and $k_{22}(t) = k_{12}(t) = k_2(t)$, it is possible to show that

$$\frac{d}{dt}\left(\int_0^\infty \rho_1(x,t)dx + \mathcal{C}_1(t)\right) = -\frac{d}{dt}\left(\int_0^\infty \rho_2(x,t)dx + \mathcal{C}_2(t)\right), \qquad (2.20)$$

which implies in

$$\mathcal{C}_1(t) + \mathcal{C}_2(t) + \int_0^\infty \rho_2(x,t)dx + \int_0^\infty \rho_1(x,t)dx = \text{constant}. \qquad (2.21)$$

Equation (2.21) is a direct consequence of the conservation of the total number of particles present in the system. Equation (2.20) also shows that the mass (number of particles) variation of the specie 1 is connected to the variations of the specie 2. In particular, the negative signal shows that the variation of the particles related to one specie (gain or loss) produces an opposite variation on the other specie. The set of Eqs. (2.11)–(2.21) covers a general scenario from which some specific situations can be investigated.

The previous equations represent a general scenario from which some particular cases, solutions, can be investigated. In this sense, we consider the case characterized by the initial conditions: $\rho_1(x,0) = \varphi_1(x)$ (with $\int_0^\infty dx\rho_1(x,0) = 1$), $\rho_2(x,0) = \varphi_2(x)$ (with $\int_0^\infty dx\rho_2(x,0) = 1$) and $\mathcal{C}_1(0) = \mathcal{C}_2(0) = 0$. The choice for the initial conditions imply that the particles are initially in the bulk. The equations previously presented can be faced by using the Laplace transform and the Green function approach in order to simplify the calculations. By applying the Laplace transform in Eqs. (2.1) and (2.2), we obtain that

$$\bar{\mathcal{K}}_1(s)\frac{\partial^2}{\partial x^2}\bar{\rho}_1(x,s) = s\bar{\rho}_1(x,t) - \rho_1(x,0), \qquad (2.22)$$

and

$$\bar{\mathcal{K}}_2(s)\frac{\partial^2}{\partial x^2}\bar{\rho}_2(x,s) = s\bar{\rho}_2(x,s) - \rho_2(x,0) \qquad (2.23)$$

with $\bar{\mathcal{K}}_1(s) = s\bar{\Phi}_1(s)\mathcal{K}_1$ and $\bar{\mathcal{K}}_2(s) = s\bar{\Phi}_2(s)\mathcal{K}_2$. Following, in the Laplace domain, the boundary conditions can be written as

$$\bar{\mathcal{K}}_1(s)\frac{\partial}{\partial x}\bar{\rho}_1(x,s)\Big|_{x=0} = \left(s\bar{\kappa}_1(s) + \bar{k}_{11}(s)\right)\bar{\rho}_1(0,s) - \bar{k}_{12}(s)\bar{\rho}_2(0,s), \quad (2.24)$$

$$\bar{\mathcal{K}}_2(s)\frac{\partial}{\partial x}\bar{\rho}_2(x,s)\Big|_{x=0} = \left(s\bar{\kappa}_2(s) + \bar{k}_{22}(s)\right)\bar{\rho}_2(0,s) - \bar{k}_{21}(s)\bar{\rho}_1(0,s), \quad (2.25)$$

$\partial_x\bar{\rho}_1(\infty,s) = 0$, and $\partial_x\bar{\rho}_2(\infty,s) = 0$. By using the Green function approach, the solutions for these equations can be found, and they are given by

$$\bar{\rho}_1(x,s) = -\int_0^\infty \mathcal{G}_1(x,x';s)\varphi_1(x')dx' + \mathcal{G}_1(x,0;s)\left(\omega_1(s)\rho_1(0,s) - k_{12}(s)\rho_2(0,s)\right) \quad (2.26)$$

and

$$\bar{\rho}_2(x,s) = -\int_0^\infty \mathcal{G}_2(x,x';s)\varphi_2(x')dx' + \mathcal{G}_2(x,0;s)\left(\omega_2(s)\rho_2(0,s) - k_{21}(s)\rho_1(0,s)\right), \quad (2.27)$$

with $\bar{\omega}_{1(2)}(s) = s\bar{\kappa}_{1(2)}(s) + \bar{k}_{11(22)}(s)$.

The Green functions, related to each specie of particles, are obtained from the following equation:

$$\bar{\mathcal{K}}_{1(2)}(s)\frac{\partial^2}{\partial x^2}\bar{\mathcal{G}}_{1(2)}(x,x';s) - s\bar{\mathcal{G}}_{1(2)}(x,x';s) = \delta(x-x') \quad (2.28)$$

subjected to the boundary conditions $\partial_x\bar{\mathcal{G}}_{1(2)}(0,x';s) = 0$ and $\partial_x\bar{\mathcal{G}}_{1(2)}(\infty,x';s) = 0$. In particular, by performing some calculations, we can show that the Green functions are given by

$$\bar{\mathcal{G}}_{1(2)}(x,x';s) = -\frac{1}{2\sqrt{s\bar{\mathcal{K}}_{1(2)}(s)}} \times \left(e^{-\sqrt{s/\bar{\mathcal{K}}_{1(2)}(s)}|x-x'|} + e^{-\sqrt{s/\bar{\mathcal{K}}_{1(2)}(s)}|x+x'|}\right). \quad (2.29)$$

In Eqs. (2.26) and (2.27), the first term promotes the spreading of the initial condition and the other terms represent the influence of the surface on the diffusive process. From these equations, after some calculations, we can show that

$$\bar{\rho}_1(0,s) = -\bar{\Lambda}(s)\left[\bar{\Phi}_2(s)\bar{\mathcal{I}}_1(s) + \bar{k}_{12}(s)\bar{\mathcal{I}}_2(s)/\left(\sqrt{s\bar{\mathcal{K}}_1(s)}\right)\right] \quad (2.30)$$

and

$$\bar{\rho}_2(0,s) = -\bar{\Lambda}(s)\left[\bar{\Phi}_1(s)\bar{\mathcal{I}}_2(s) + \bar{k}_{21}(s)\bar{\mathcal{I}}_1(s)/\left(\sqrt{s\bar{\mathcal{K}}_2(s)}\right)\right], \quad (2.31)$$

where

$$\bar{\Lambda}(s) = 1 \bigg/ \left[\bar{\Phi}_1(s)\bar{\Phi}_2(s) - \bar{k}_{21}(s)\bar{k}_{12}(s)\Big/\left(s\sqrt{\bar{\mathcal{K}}_1(s)\bar{\mathcal{K}}_2(s)}\right)\right], \tag{2.32}$$

$\bar{\Phi}_{1(2)}(s) = 1 + \bar{\omega}_{1(2)}(s)/\left(\sqrt{s\bar{\mathcal{K}}_{1(2)}(s)}\right)$ and $\bar{\mathcal{I}}_{1(2)}(s) = \int_0^\infty dx'\bar{\mathcal{G}}_{1(2)}(0,x';s)$ $\varphi_{1(2)}(x')$. These equations also enable us to obtain the survival probability ($S_{1(2)}(t) = \int_0^\infty dx\rho_{1(2)}(x,t)$), which is related to the quantity of substance present in the bulk. By performing some calculations, we obtain, for each specie, that

$$\bar{S}_1(s) = \left[1 - \bar{\omega}_1(s)\bar{\rho}_1(0,s) + \bar{k}_{12}(s)\bar{\rho}_2(0,s)\right]/\, s, \tag{2.33}$$

$$\bar{S}_2(s) = \left[1 - \bar{\omega}_2(s)\bar{\rho}_2(0,s) + \bar{k}_{21}(s)\bar{\rho}_1(0,s)\right]/\, s. \tag{2.34}$$

Now, let us analyze some cases which can be obtained from the previous calculations. One of them is obtained by considering that $\kappa_1(t) = 0$ and $\kappa_2(t) = 0$, i.e., absence of adsorption process by the surface, with $k_{11}(t) = k_{21}(t) = k_1(t)$ and $k_{22}(t) = k_{12}(t) = k_2(t)$. This case implies that the particles 1 and 2 are absorbed by the surface to promote the reaction process $1 \rightleftarrows 2$, where each substance promotes the formation of the other which is released to the bulk. This reaction process on the surface is typical of a reversible reaction. For this case, the distributions related to each specie can be written as follows:

$$\bar{\rho}_1(x,s) = -\int_0^\infty dx'\bar{\mathcal{G}}_1(x,x';s)\varphi_1(x') - \frac{\sqrt{s}\bar{\mathcal{G}}_1(x,0,s)}{\sqrt{s} + \bar{k}_1(s)/\sqrt{\bar{\mathcal{K}}_1(s)} + \bar{k}_2(s)/\sqrt{\bar{\mathcal{K}}_2(s)}}\left(\bar{k}_1(s)\bar{\mathcal{I}}_1(s) - \bar{k}_2(s)\bar{\mathcal{I}}_2(s)\right), \tag{2.35}$$

$$\bar{\rho}_2(x,s) = -\int_0^\infty dx'\bar{\mathcal{G}}_2(x,x';s)\bar{\varphi}_2(x') - \frac{\sqrt{s}\bar{\mathcal{G}}_2(x,0,s)}{\sqrt{s} + \bar{k}_1(s)/\sqrt{\bar{\mathcal{K}}_1(s)} + \bar{k}_2(s)/\sqrt{\bar{\mathcal{K}}_2(s)}}\left(\bar{k}_2(s)\bar{\mathcal{I}}_2(s) - \bar{k}_1(s)\bar{\mathcal{I}}_1(s)\right). \tag{2.36}$$

The survival probability, related to the quantity of each species of particle present in the bulk, for this case is given by

$$\bar{S}_1(s) = \frac{1}{s}\left[1 + \frac{\sqrt{s}\left(\bar{k}_1(s)\bar{\mathcal{I}}_1(s) - \bar{k}_2(s)\bar{\mathcal{I}}_2(s)\right)}{\sqrt{s} + \bar{k}_1(s)/\sqrt{\bar{\mathcal{K}}_1(s)} + \bar{k}_2(s)/\sqrt{\bar{\mathcal{K}}_2(s)}}\right], \tag{2.37}$$

$$\bar{S}_2(s) = \frac{1}{s}\left[1 + \frac{\sqrt{s}\left(\bar{k}_2(s)\bar{\mathcal{I}}_2(s) - \bar{k}_1(s)\bar{\mathcal{I}}_1(s)\right)}{\sqrt{s} + \bar{k}_1(s)/\sqrt{\bar{\mathcal{K}}_1(s)} + \bar{k}_2(s)/\sqrt{\bar{\mathcal{K}}_2(s)}}\right], \tag{2.38}$$

which implies in $\mathcal{S}_1(t)+\mathcal{S}_2(t) = constant$. Equations (2.43) and (2.38) for the particular case $\bar{\mathcal{K}}_{1(2)}(s) = \bar{\mathcal{K}}(s)$ with $\varphi_{1(2)}(x) = \delta(x-x')$ can be simplified to

$$\bar{\mathcal{S}}_1(s) = \frac{1}{s} - \frac{1}{s}\left(\frac{\bar{k}_1(s)e^{-\sqrt{s/\bar{\mathcal{K}}(s)}|x'|}}{\sqrt{s\bar{\mathcal{K}}(s)}+\bar{k}_1(s)+\bar{k}_2(s)} - \frac{\bar{k}_2(s)e^{-\sqrt{s/\bar{\mathcal{K}}(s)}|x'|}}{\sqrt{s\bar{\mathcal{K}}(s)}+\bar{k}_1(s)+\bar{k}_2(s)}\right), \tag{2.39}$$

$$\bar{\mathcal{S}}_2(s) = \frac{1}{s} - \frac{1}{s}\left(\frac{\bar{k}_2(s)e^{-\sqrt{s/\bar{\mathcal{K}}(s)}|x'|}}{\sqrt{s\bar{\mathcal{K}}(s)}+\bar{k}_1(s)+\bar{k}_2(s)} - \frac{\bar{k}_1(s)e^{-\sqrt{s/\bar{\mathcal{K}}(s)}|x'|}}{\sqrt{s\bar{\mathcal{K}}(s)}+\bar{k}_1(s)+\bar{k}_2(s)}\right). \tag{2.40}$$

For the general results represented by Eqs. (2.35)–(2.38), we consider different time dependencies for $\bar{\mathcal{K}}(s)$, $\bar{k}_1(s)$ and $\bar{k}_2(s)$ in to perform the inverse Laplace transform. The first choice to be considered is $\bar{\mathcal{K}}_{1(2)}(s) = \mathcal{K}_{1(2)}/s^{\alpha_{1(2)}}$ $(\mathcal{K}_{1(2)}(t) = \mathcal{K}_{1(2)}t^{\alpha_{1(2)}-1}/\Gamma(\alpha_{1(2)})$, $\bar{k}_1(s) = k_1$ $(k_1(t) = k_1\delta(t))$ and $\bar{k}_2(s) = k_2$ $(k_2(t) = k_2\delta(t))$. It is worth mentioning that this choice for the time dependence for diffusion coefficient corresponds to the Riemann–Liouville fractional time operator. For this case, the Green function is given by

$$\mathcal{G}^{RL}_{1(2)}(x,x';t) = -\frac{1}{\sqrt{4\mathcal{K}_{1(2)}t^{\alpha_{1(2)}}}}\left(\mathrm{H}^{1,0}_{1,1}\left[\frac{|x-x'|}{\sqrt{\mathcal{K}_{1(2)}t^{\alpha_{1(2)}}}}\left|\begin{matrix}\left(1-\frac{\alpha_{1(2)}}{2},\frac{\alpha_{1(2)}}{2}\right)\\(0,\,1)\end{matrix}\right.\right] + \mathrm{H}^{1,0}_{1,1}\left[\frac{|x+x'|}{\sqrt{\mathcal{K}_{1(2)}t^{\alpha_{1(2)}}}}\left|\begin{matrix}\left(1-\frac{\alpha_{1(2)}}{2},\frac{\alpha_{1(2)}}{2}\right)\\(0,\,1)\end{matrix}\right.\right]\right) \tag{2.41}$$

where $\mathrm{H}^{m,n}_{p,q}\left[x\left|\begin{matrix}(a_p,A_p)\\(b_q\ B_q)\end{matrix}\right.\right]$ is the Fox H function [46]. By performing some calculations, it is possible to show that the asymptotic behaviour for this case is given by

$$\mathcal{G}^{RL}_{1(2)}(x,x';t) \sim -\frac{1}{\sqrt{4(2-\alpha_{1(2)})\pi\mathcal{K}_{1(2)}t^{\alpha_{1(2)}}}}\left(\frac{\alpha}{\sqrt{4\mathcal{K}_{1(2)}t^{\alpha_{1(2)}}}}|x-x'|\right)^{\frac{\alpha_{1(2)}-1}{2-\alpha_{1(2)}}} \times\left\{\exp\left[-\tilde{\alpha}_{1(2)}\left(\frac{|x-x'|}{\sqrt{\mathcal{K}_{1(2)}t^{\alpha_{1(2)}}}}\right)^{\frac{2}{2-\alpha_{1(2)}}}\right] + \exp\left[-\tilde{\alpha}_{1(2)}\left(\frac{|x+x'|}{\sqrt{\mathcal{K}_{1(2)}t^{\alpha_{1(2)}}}}\right)^{\frac{2}{2-\alpha_{1(2)}}}\right]\right\}, \tag{2.42}$$

where $\tilde{\alpha}_{1(2)} = [(2-\alpha_{1(2)})/2]\left(\alpha_{1(2)}/2\right)^{\alpha_{1(2)}/(2-\alpha_{1(2)})}$. Equation (2.42) shows that the asymptotic behaviour is essentially characterized by a stretched-like exponential behaviour and for $\alpha = 1$ the usual result is recovered. In this scenario, Eqs. (2.39) and (2.40) are given by

$$\mathcal{S}_1(t) = 1 - (k_1 - k_2) \int_0^t \frac{dt'}{\sqrt{\mathcal{K} t'^{1+\gamma}}} E_\alpha \left(-\frac{k_t}{\sqrt{\mathcal{K}}} (t-t')^\gamma \right)$$

$$\mathrm{H}_{1,1}^{1,0} \left[\frac{|x'|}{\sqrt{\mathcal{K} t'^\alpha}} \middle| \begin{matrix} (\sigma, \frac{\alpha}{2}) \\ (0,\, 1) \end{matrix} \right], \tag{2.43}$$

and $\mathcal{S}_2(t) = 2 - \mathcal{S}_1(t)$ with $k_t = k_1 + k_2$ and $\sigma = (1-\alpha)/2$.

For the Caputo–Fabrizio operator, where

$$\bar{\mathcal{K}}_{1(2)}(s) = \frac{\tilde{\mathcal{K}}_{1(2)}}{s^{\alpha_{1(2)}} + \bar{\alpha}_{1(2)}}, \tag{2.44}$$

with $\tilde{\mathcal{K}}_{1(2)} = \mathcal{R}_{1(2)} \mathcal{K}_{1(2)}$ and $\mathcal{R}_{1(2)} = \mathcal{R}(\alpha_{1(2)})$, the Green function is given by

$$\begin{aligned} \mathcal{G}_{1(2)}^{CF}(x,x';t) = &-\frac{1}{\sqrt{4\tilde{\mathcal{K}}_{1(2)} t}} \left(e^{-\frac{(x-x')^2}{4\tilde{\mathcal{K}}_{1(2)} t}} + e^{-\frac{(x+x')^2}{4\tilde{\mathcal{K}}_{1(2)} t}} \right) \\ &- \int_0^t dt' \frac{e^{-\alpha_{1(2)} t'}}{\sqrt{4\tilde{\mathcal{K}}_{1(2)} t'}} \left(e^{-\frac{(x-x')^2}{4\tilde{\mathcal{K}}_{1(2)} t'}} + e^{-\frac{(x+x')^2}{4\tilde{\mathcal{K}}_{1(2)} t'}} \right). \end{aligned} \tag{2.45}$$

It is worth mentioning that the previous equation obtained for the Caputo–Fabrizio fractional time operator presents for long times a time-independent behaviour. In particular, it is possible to show, in the asymptotic limit of $t \to \infty$, that

$$\begin{aligned} \mathcal{G}_{1(2)}^{CF}(x,x';t) \to \mathcal{G}_{st,1(2)}^{CF}(x,x') = &-\frac{1}{2\sqrt{\tilde{\mathcal{K}}_{1(2)}}} \\ &\times \left(e^{-\sqrt{\frac{\overline{\alpha}_{1(2)}}{\tilde{\mathcal{K}}_{1(2)}}}|x-x'|} + e^{-\sqrt{\frac{\overline{\alpha}_{1(2)}}{\tilde{\mathcal{K}}_{1(2)}}}|x+x'|} \right). \end{aligned} \tag{2.46}$$

Thus, the relaxation process exhibited by each one of these Green functions shows the main difference among these fractional differential operator, when employed to describe a diffusive process. In fact, the fractional Caputo–Fabrizio presents, as shown earlier, a steady state, in contrast to the others considered here. This feature may be related to the resetting process associated with this operator as discussed in Ref. [51].

The Atangana–Baleanu fractional differential operator considered here introduces different diffusive regimes on the spreading of the system as discussed in Ref. [51], where this fractional operator was related to a fractional time derivative of distributed order. For this case, in the Laplace domain, we have that

$$\bar{\mathcal{K}}_{1(2)}(s) = \tilde{\mathcal{K}}'_{1(2)} \frac{s^{\alpha_{1(2)}}}{s^{\alpha_{1(2)}} + \bar{\alpha}_{1(2)}}, \tag{2.47}$$

where $\tilde{\mathcal{K}}'_{1(2)} = \mathcal{R}_{1(2)}\mathcal{K}_{1(2)}$ and $\mathcal{R}_{1(2)} = \mathcal{R}(\alpha_{1(2)})$ and, after some calculations, the Green function is given by

$$\mathcal{G}_{\mathcal{AB}}(x,x';t) = -\sum_{n=0}^{\infty} \frac{(-\bar{\alpha}_{1(2)})^n}{\Gamma(1+n)} t^{n\alpha_{1(2)}} \left(\mathcal{G}^{(-)}_{\mathcal{AB}}(x,x',t) + \mathcal{G}^{(+)}_{\mathcal{AB}}(x,x',t)\right) \tag{2.48}$$

with

$$\mathcal{G}^{(\pm)}_{\mathcal{AB}}(x,x',t) = \frac{1}{\sqrt{4\tilde{\mathcal{K}}'_{1(2)} t^{\alpha_{1(2)}}}} \left(\mathrm{H}^{1,0}_{1,1}\left[\frac{|x \pm x'|}{\sqrt{\tilde{\mathcal{K}}'_{1(2)} t^{\alpha_{1(2)}}}} \middle| \begin{matrix} (\frac{1}{2}+n\alpha_{1(2)}, \frac{1}{2}) \\ (0,\,1) \end{matrix}\right] \right.$$
$$\left. + t^{\alpha_{1(2)}-1}\, \mathrm{H}^{1,0}_{1,1}\left[\frac{|x \pm x'|}{\sqrt{\tilde{\mathcal{K}}'_{1(2)} t^{\alpha_{1(2)}}}} \middle| \begin{matrix} (\frac{1}{2}+(1+n)\alpha_{1(2)}, \frac{1}{2}) \\ (0,\,1) \end{matrix}\right] \right). \tag{2.49}$$

Figure 2.1 illustrates the behaviour for the Green function by taking into account the power-law kernel related to the Riemann–Liouville fractional operator. In Figure 2.2, we consider an exponential kernel for the diffusion equations. In contrast to the Riemann–Liouville, this kernel is non-singular

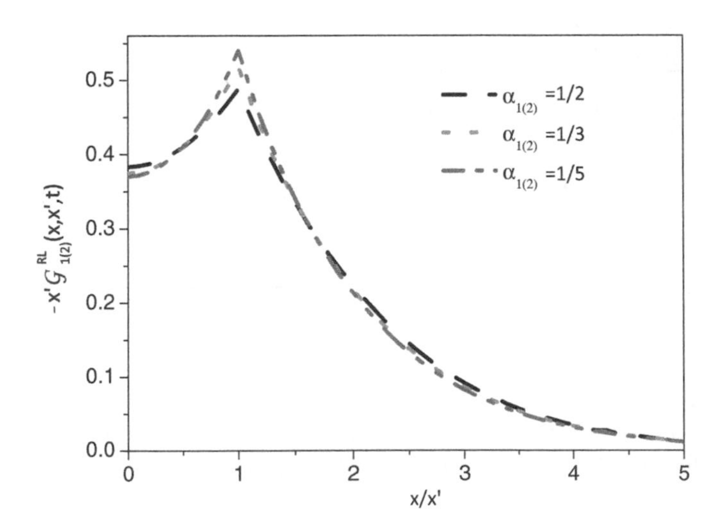

FIGURE 2.1
This figure illustrates the behaviour of Eq. (2.41) for different values of $\alpha_{1(2)}$. We consider, for simplicity, $\mathcal{K}_{1(2)}t^{\alpha_{1(2)}}/x'^2 = 1$.

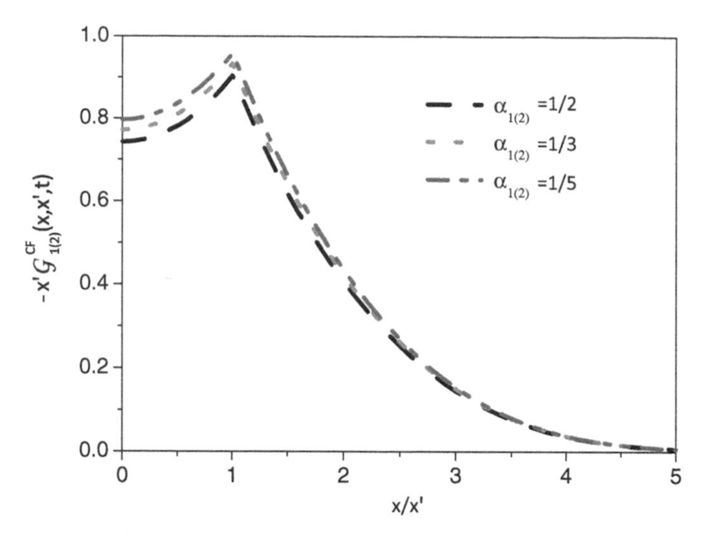

FIGURE 2.2
This figure illustrates the behaviour of Eq. (2.41) for different values of α. We consider, for simplicity, $\mathcal{K}_{1(2)}t/x'^2 = 1$.

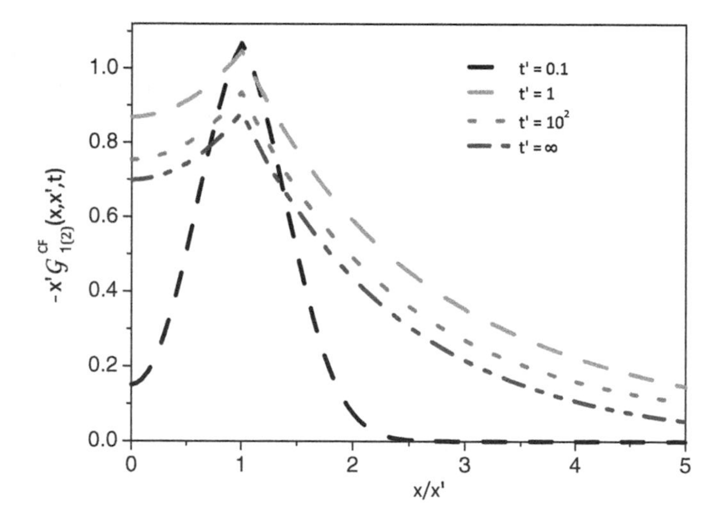

FIGURE 2.3
This figure illustrates the behaviour of Eq. (2.45) for different values of t. We consider, for simplicity, $\mathcal{K}_{1(2)}t/x'^2 = t'$ and $\alpha_{1(2)} = 1/2$.

at the origin and leads us to a stationary solution for long times as shown in Figure 2.3. Figure 2.4 compares the Green function obtained with these operators with the usual one.

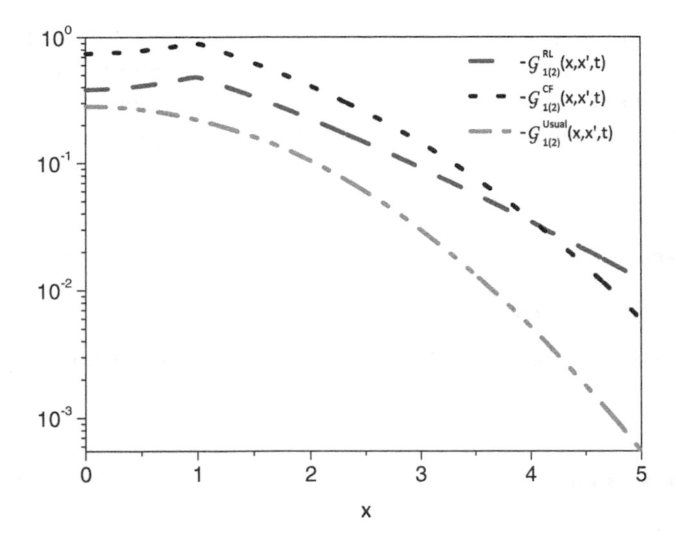

FIGURE 2.4
This figure compares the behaviour of Eqs. (2.41) and (2.45) with the usual one for $\alpha_{1(2)} = 1/2$. We consider, for simplicity, $\mathcal{K}_{1(2)} = 1$, and $x' = 1$.

2.3 Discussion and Conclusions

We have investigated a diffusion process subject to surface effects that may adsorb, desorb and/or absorb. The particles in the bulk are subjected to diffusion-like equations that extend the usual one to a broad class of situations, in particular the fractional diffusion equations. In the case of fractional diffusion equations, they can be related to fractional diffusion equations with singular or non-singular kernels. For the surface effects, we consider boundary conditions that are able to describe processes such as adsorption, desorption, absorption and reaction. For the adsorption–desorption process, we consider them described by Eqs. (2.11) and (2.12). These equations may be related to a large class of processes depending on the choice performed for $\kappa_i(t)$. In particular, we may have Debye or non-Debye relations, where the last ones may be related to memory effects present in the system. The particles absorbed may promote a reaction process and the formation of other compounds. This process of reaction is governed by the kernels present in the boundary conditions, i.e., $k_{11}(t)$, $k_{12}(t)$, $k_{21}(t)$ and $k_{22}(t)$. In particular, depending on the choice of these kernels the reaction process manifested by the surface can be irreversible ($1 \rightarrow 2$ or $1 \leftarrow 2$) or reversible ($1 \rightleftarrows 2$). Following, we have worked out a scenario characterized by a reversible reaction process by taking different fractional operators into account. The Green function is illustrated in Figures 2.1 and 2.2. In Figure 2.3, we have shown that the exponential kernel leads us to a stationary state, in contrast to the Riemann–Liouville fractional time operator. Figure 2.4 compares the different differential operators and the usual. Finally, we hope that the results presented here may be useful in contexts where anomalous diffusion and effect on surfaces are present.

Acknowledgement

The authors are thankful to CNPq (Brazilian agencies) for financial support.

References

[1] Strizhak, P. E. (2004). Macrokinetics of chemical processes on corous catalysts having regard to anomalous diffusion. *Theoretical and Experimental Chemistry*, 40, 203–208.

[2] Avnir, D., *The Fractal Approach to Heterogeneous Chemistry*, Wiley-Interscience, New York, 1990.

[3] Pekalski, A. & Weron, K. S., *Anomalous Diffusion: From Basics to Applications, Lecture Notes in Physics* (vol. 519), Springer, Berlin, 1998.

[4] Lenzi, E. K., Lenzi, M. K., Zola, R. S., Ribeiro, H. V., Zola, F. C., Evangelista, L. R. & Gonçalves, G. (2014). Reaction on a solid surface supplied by an anomalous mass transfer source. *Physica A*, 410, 399–406.

[5] Mendez, V., Campos, D. & Bartumeus, F., *Stochastic Foundations in Movement Ecology*, Springer, Heidelberg, 2014.

[6] Ben-Avraham, D. & Havlin, S., *Diffusion and Reactions in Fractals and Disordered Systems*, Cambridge University Press, Cambridge, 2005.

[7] Fogler, H. S., *Elements of Chemical Reaction Engineering*, Prentice Hall, NJ, 1999.

[8] Fibich, G., Gannot, I., Hammer, A. & Schochet, S. (2006). Chemical kinetics on surfaces: a singular limit of a reaction–diffusion system. *SIAM Journal on Mathematical Analysis*, 38, 1371–1388.

[9] Bressloff, P. C., *Stochastic Processes in Cell Biology*, Springer, Heidelberg, 2014.

[10] Guimarães, V. G., Ribeiro, H. V., Li, Q., Evangelista, L. R., Lenzi, E. K. & Zola, R. S. (2015). Unusual diffusing regimes caused by different adsorbing surfaces. *Soft Matter*, 11, 1658–1666.

[11] Lima, D. F. B., Zanella, F. A., Lenzi, M. K. & Ndiaye, P. M., Modeling and Simulation of Water Gas Shift Reactor: An Industrial Case, *Petrochemicals*, Dr Vivek Patel (Ed.), InTech, 2012.

[12] Friesen, V. C., Leitoles, D. P., Gonalves, G., Lenzi, E. K. & Lenzi, M. K. (2015). Modeling heavy metal sorption kinetics using fractional calculus. *Mathematical Problem in Engineering*, 2015, 549562.

[13] Lenzi, E. K., Ribeiro, H. V., Tateishi, A. A., Zola, R. S. & Evangelista, L. R. (2016). Anomalous diffusion and transport in heterogeneous systems separated by a membrane transport of particles. *Proceedings of the Royal Society A: Mathematical, Physical and Engineering Sciences*, 472, 0502.

[14] Lenzi, E. K., Fernandes, P. R. G., Petrucci, T., Mukai, H. & Ribeiro, H. V. (2011). Anomalous–diffusion approach applied to the electrical response of water. *Physical Review E*, 84, 041128.

[15] Ciuchi, F., Mazzulla, A., Scaramuzza, N., Lenzi, E. K. & Evangelista, L. R. (2012). Fractional diffusion equation and the electrical impedance: experimental evidence in liquid–crystalline cells. *The Journal of Physical Chemistry C*, 116, 8773–8777.

[16] Crank, J., *The Mathematics of Diffusion*, Oxford Science Publications, England, 1975.

[17] Giona, M. & Roman, H. E. (1992). A theory of transport phenomena in disordered systems. *Chemical Engineering Journal*, 49, 1–10.

[18] Snopok, B. A. (2014). Nonexponential kinetic of surface chemical reactions. *Theoretical and Experimental Chemistry*, 50, 67–95.

[19] Klafter, J. & Sokolov, I. M., *First Steps in Random Walks: From Tools to Applications*, Oxford University Press, Oxford, 2011.

[20] Lenzi, E. K., Ribeiro, H. V., Martins, J., Lenzi, M. K., Lenzi, G. G. & Specchia, S. (2011). Non–markovian diffusion equation and diffusion in a porous catalyst. *Chemical Engineering Journal*, 172, 1083–1087.

[21] Hilfer, R., Metzler, R., Blumen, A. & Klafter, J. (2002). Strange kinetics. *Chemical Physics*, 284, 1–542.

[22] Metzler, R. & Klafter, J. (2000). The random walk's guide to anomalous diffusion: a fractional dynamics approach. *Physics Report*, 339, 1–77.

[23] Metzler, R. & Klafter, J. (2004). The restaurant at the end of the random walk: recent developments in the description of anomalous transport by fractional dynamics. *Journal of Physics A: Mathematical and General*, 37, R161–R208.

[24] Bressloff, P. C. & Newby, J. M. (2013). Stochastic models of intracellular transport. *Review Modern Physics*, 85, 135–196.

[25] Lucena, L. S., da Silva, L. R., Evangelista, L. R., Lenzi, M. K., Rossato, R. & Lenzi, E. K. (2008). Solutions for a fractional diffusion equation with spherical symmetry using green function approach. *Chemical Physics*, 344, 90–94.

[26] Giona, M. & Giustiniani, M. (1996). Adsorption kinetics on fractal surfaces. *The Journal of Physical Chemistry*, 100, 16690–16699.

[27] Yang, X. J. & Baleanu, D. (2013). Fractal heat conduction problem solved by local fractional variation iteration method. *Thermal Science*, 17, 625–628.

[28] Yang, X. J., Baleanu, D. & Srivastava, H. M. (2015). Local fractional similarity solution for the diffusion equation defined on Cantor sets. *Applied Mathematics Letters*, 47, 54–60.

[29] Hristov, J. (2010). Heat–balance integral to fractional (half-time) heat diffusion sub-model. *Thermal Science*, 14, 291–316.

[30] Robson, A., Burrage, K. & Leake, M. C. (2013). Inferring diffusion in single live cells at the single–molecule level. *Philosophical Transactions of the Royal Society B: Biological Sciences*, 368, 1471–2970.

[31] Caputo, M., Cametti, C. & Ruggero, V. (2008). Time and spatial concentration profile inside a membrane by means of a memory formalism. *Physica A*, 387, 2010–2018.

[32] Caputo, M. & Cametti, C. (2009). The memory formalism in the diffusion of drugs through skin membrane. *Journal of Physics D: Applied Physics*, 42, 12505.

[33] Kosztolowicz, T., Dworecki, K. & Lewandowska, K. D. (2012). Subdiffusion in a system with thin membranes. *Physical Review E*, 86, 021123.

[34] Santoro, P. A., de Paula, J. L., Lenzi, E. K. & Evangelista, L. R. (2011). Anomalous diffusion governed by a fractional diffusion equation and the electrical response of an electrolytic cell. *The Journal of Chemical Physics*, 135, 114704.

[35] Caspi, A., Granek, R. & Elbaum, M. (2000). Enhanced diffusion in active intracellular transport. *Physical Review Letters*, 85, 5655–5658.

[36] Sokolov, I. M. (2012). Models of anomalous diffusion in crowded environments. *Soft Matter*, 8, 9043–9052.

[37] Weigel, A. V., Simon, B., Tamkun, M. M. & Krapf, D. (2011). Ergodic and nonergodic processes coexist in the plasma membrane as observed by single-molecule tracking. *Proceedings of the National Academy of Science of the United States of America*, 108, 6439–6443.

[38] Jaing, H., Cheng, Y., Tuan, L., An, F. & Jin, K. (2013). A fractal theory based fractional diffusion model used for the fast desorption process of methane in coal. *Chaos*, 23, 033111.

[39] Kosztolowicz, T. (2015). Random walk model of subdiffusion in a system with a thin membrane. *Physical Review E*, 91, 022102.

[40] Podlubny, I., *Fractional Differential Equations*, Academic Press, San Diego, CA, 1999.

[41] Caputo, M. & Fabrizio, M. (2015). A new definition of fractional derivative without singular kernel. *Progress in Fractional Differentiation and Applications*, 1, 73–85.

[42] Atangana, A. & Baleanu, D. (2016). New fractional derivative with non-local and non-singular kernel. *Thermal Science*, 20, 763–769.

[43] Gómez-Aguilar J. F. & Atangana, A. (2017). Fractional Hunter-Saxton equation involving partial operators with bi-order in Riemann-Liouville and Liouville-Caputo sense. *The European Physical Journal Plus*, 132, 100.

[44] Hristov, J. (2016). Transient heat diffusion with a non-singular fading memory: From the Cattaneo constitutive equation with Jeffrey's kernel to the Caputo-Fabrizio time-fractional derivative. *Thermal Science*, 20, 757–762.

[45] Gómez-Aguilar, J. F. (2017). Space–time fractional diffusion equation using a derivative with nonsingular and regular kernel. *Physica A*, 465, 562–572.

[46] Mathai, A. M., Saxena, R. K. & Haubold, H. J., *The H-Function: Theory and Applications*, Springer, New York, 2009.

[47] Zola, R. S., Lenzi, E. K., Evangelista, L. R., & Barbero, G. (2007). Memory effect in the adsorption phenomena of neutral particles. *Physical Review E*, 75, 042601.

[48] Lenzi, E. K., Yednak, C. A. R. & Evangelista, L. R. (2010). Non-Markovian diffusion and the adsorption–desorption process. *Physical Review E*, 81, 011116.

[49] de Paula, J. L., Santoro, P. A., Zola, R. S., Lenzi, E. K., Evangelista, L. R., Ciuchi, F., Mazzulla, A. & Scaramuzza, N. (2012). Non–Debye relaxation in the dielectric response of nematic liquid crystals: surface and memory effects in the adsorption–desorption process of ionic impurities. *Physical Review E*, 86, 051705.

[50] Kopelman, R. (1988). Fractal reaction kinetics. *Science*, 241, 1620–1626.

[51] Tateishi, A. A., Ribeiro H. V., & Lenzi, E. K. (2017). The role of fractional time–derivative operators on anomalous diffusion. *Frontiers in Physics*, 5, 52.

3

An Efficient Computational Method for Non-Linear Fractional Lienard Equation Arising in Oscillating Circuits

Harendra Singh

Post Graduate College, Ghazipur

CONTENTS

3.1 Introduction

The standard Lienard equation is considered as a generalization of the spring-mass system or damped pendulum equation. This equation played an important role in the development of radio and vacuum tube technology, because it can be applied to describe the oscillating circuits. The Lienard equation is a highly non-linear second-order differential equation that was given by Lienard [1] and is presented as

$$x'' + b_1(x)x' + b_2(x) = b_3(t), \tag{3.1}$$

where $b_1(x)x'$ is the damping force, $b_2(x)$ is the restoring force and $b_3(t)$ is the external force. This equation is used as a non-linear model in many physical phenomena for different choices of $b_1(x)$, $b_2(x)$ and $b_3(t)$. In particular, if we take $b_1(x) = \varepsilon(x^2 - 1)$, $b_2(x) = x$ and $b_3(t) = 0$ in Eq. (3.1), then it becomes the Van der Pol equation, which act as a non-linear model of electronic oscillations [2,3].

It is very difficult to find the exact solution of these equations by usual ways [4]. Kong studied the following form of Lienard equation [5]

$$x'' + px' + qx^3 + rx^5 = 0, \tag{3.2}$$

where p, q and r are real constants.

Nowadays, fractional calculus has become a very important and useful branch of mathematics. The importance of fractional calculus arises because an accurate modelling of a physical phenomenon that has dependence at both present and preceding times can be accurately understood with fractional calculus. Moreover, the real-life applications of fractional calculus in engineering and sciences include fluid dynamics [6], signal processing [7], chemistry [8], viscoelasticity [9–11], bioengineering [12] and control theory [13]. Due to real-life applications and for the accurate modelling of physical phenomenon, it is necessary to replace integer order derivatives to fractional order derivatives. In the present article, we will consider the more general form of the Lienard equation given by

$$x^{\alpha} + px' + qx^3 + rx^5 = 0, \ 1 < \alpha \leq 2, \ t \in [0, 1], \tag{3.3}$$

with initial conditions:

$$x(0) = \varsigma, \quad x'(0) = \tau, \tag{3.4}$$

where ς and τ are real constants.

There are some analytical and numerical methods to solve the Lienard equation with integer order. The pioneer approach for solving the Lienard equation originates by Kong [5], who gave an exact solution of this equation for some particular cases. In [14], Feng obtained an exact solution of this equation for some particular choice of real constants, and these results were the generalization of Kong results. In 2008, the authors proposed a method based on variational iteration method to obtain approximate solutions of Lienard equation [15]. In 2011, variational homotopy perturbation method was proposed to solve Lienard equation [16]. Recently in 2017, the authors proposed a numerical method based on homotopy analysis transform method to solve fractional Lienard equation and showed the uniqueness and existence of solutions [17].

In this article, we present a computational method to solve fractional Lienard equation. The proposed method is based on the applications of operational matrix of differentiations method and collocation method for Legendre scaling functions. Operational matrix method is used to solve many problems in differential calculus, see [18–25]. Using operational matrix of differentiations and collocation method, we convert the FLE into a system of non-linear algebraic equations whose solution gives approximate solution to FLE. The solution is discussed for different fractional orders. The results are compared with the exact solution and are presented in the form of tables.

3.2 Preliminaries

In this article, we have used fractional derivative in Liouville–Caputo sense.

Definition 3.1 The Liouville–Caputo fractional derivative of order $\alpha \geq 0$ is given as

$$(D^{\alpha}h)(t) = \begin{cases} \frac{1}{\Gamma(l-\alpha)} \int_0^t (t-x)^{l-\alpha-1} \frac{\mathrm{d}^l}{\mathrm{d}x^l} h(x)\mathrm{d}x, & l-1 < \alpha < l. \\ \frac{\mathrm{d}^l}{\mathrm{d}t^l} h(t), \quad \alpha = l \in N. \end{cases}$$

The Legendre scaling functions $\{\psi_i(t)\}$ in one dimension are defined by

$$\psi_i(t) = \begin{cases} \sqrt{(2i+1)} L_i(2t-1), & \text{for } 0 \leq t < 1. \\ 0, \quad \text{otherwise,} \end{cases}$$

where $L_i(t)$ is Legendre polynomials of order i on the interval $[-1, 1]$. These functions form an orthonormal basis for $L^2[0, 1]$, because they are constructed by normalizing the shifted Legendre polynomials. The Legendre scaling function of degree i is given by [26]

$$\psi_i(t) = (2i+1)^{\frac{1}{2}} \sum_{k=0}^{i} (-1)^{i+k} \frac{(i+k)!}{(i-k)!} \frac{t^k}{(k!)^2} \tag{3.5}$$

A function $g \in L^2[0, 1]$, with bounded second derivative $|g''(t)| \leq K$, expanded as an infinite sum of Legendre scaling function, and the series converges uniformly to the function $g(t)$,

$$g(t) = \lim_{m \to \infty} \sum_{i=0}^{m} c_i \psi_i(t), \tag{3.6}$$

where $c_i = \langle f(t), \phi_i(t) \rangle$, and $\langle .,. \rangle$ is the standard inner product on $L^2[0, 1]$.

Truncating the series at $m = n$, we get

$$g \cong \sum_{i=0}^{n} c_i \psi_i = C^T \psi(t), \tag{3.7}$$

where C and $\psi(t)$ are $(n+1) \times 1$ matrices given by

$$C = [c_0, c_1, \ldots, c_n]^T \text{ and } \psi(t) = [\psi_0(t), \psi_1(t), \ldots, \psi_n(t)]^T.$$

Theorem 3.1 *Let $\psi(t) = [\psi_0(t), \psi_1(t), \ldots, \psi_n(t)]^T$, be Legendre scaling vector and consider $\alpha > 0$, then*

$$D^{\alpha} \psi_i(t) = D^{(\alpha)} \psi(t), \tag{3.8}$$

where $D^{(\alpha)}=(\zeta(i,j))$ *is an* $(n+1)\times(n+1)$ *operational matrix of Liouville–Caputo fractional derivative of order* α *and its* (i,j)*th entry is given by*

$$\zeta(i,j)=(2i+1)^{1/2}(2j+1)^{1/2}\sum_{k=\lceil\alpha\rceil}^{i}\sum_{l=0}^{j}(-1)^{i+j+k+l}\frac{(i+k)!(j+l)!}{(i-k)!(j-l)!(k)!(l!)^2(k+l+1-\alpha)\Gamma(k+1-\alpha)}.$$

Proof: Please see [26,27].

3.3 Method of Solution

In this section, we present a computational method for the approximate solution of fractional order Lienard equation. Let us take the following approximation

$$x(t)=\sum_{i=0}^{n}c_i\psi_i(t)=C^T\psi(t). \tag{3.9}$$

Using this approximation in Eq. (3.3), we get

$$C^TD^{(\alpha)}\psi(t)+pC^TD^{(1)}\psi(t)+q\left(C^T\psi(t)\right)^3+r\left(C^T\psi(t)\right)^5=0. \tag{3.10}$$

The residual for Eq. (3.10) can be written as

$$R_n(t)=C^TD^{(\alpha)}\psi(t)+pC^TD^{(1)}\psi(t)+q\left(C^T\psi(t)\right)^3+r\left(C^T\psi(t)\right)^5, \tag{3.11}$$

where $D^{(\alpha)}$ and $D^{(1)}$ are operational matrix of differentiations of order α and 1, respectively, given by equation (3.8). Now collocate Eq. (3.11), at $n-1$ points given by $x_i=\frac{i}{n}$, $i=0,1,2,\ldots,n-2$.

Initial conditions given in Eq. (3.4) are given by

$$C^T\psi(0)=\varsigma,\quad C^TD^{(1)}\psi(0)=\tau. \tag{3.12}$$

Using collocation points in Eq. (3.11) along with the Eq. (3.12), we get a system of $n+1$ non-linear algebraic equations whose solution gives the unknown in the approximation. Using these values of unknowns in Eq. (3.9), we get an approximate solution for FLE.

3.4 Numerical Experiments and Discussion

In this section, we study the applicability and accuracy of our proposed computational method by applying it on Lienard equation with differential initial conditions. We choose the constants in Lienard equation in such a way that the exact solution is known for that case. For each case, we show how the approximate solution varies with fractional order involved in Lienard equation. In the figures, box α is denoted by alpha.

Case 1. In this case, we construct a numerical solution for Lienard equation given by Eq. (3.3) with the following initial conditions [15,17]

$$x(0) = \varsigma = \sqrt{\frac{-2p}{q}} \text{ and } x'(0) = \tau = -\frac{p\sqrt{-p}}{q\sqrt{\frac{-2p}{q}}}. \tag{3.13}$$

For this case, we take $p = -1, q = 4$ and $r = -3$. The exact solution for integer order of Eq. (3.3) with initial condition in Eq. (3.13) is given as

$$x(t) = \sqrt{\frac{-2p\left(1 + \tanh\sqrt{-p}t\right)}{q}}. \tag{3.14}$$

In Figure 3.1, we have plotted an approximate solution for different values of $\alpha = 1.7$, 1.75, 1.8, 1.85, 1.9, 1.95 and 2.

From Figure 3.1, it is observed that the solution varies continuously from fractional order to integer order. In Table 3.1, we have listed approximate

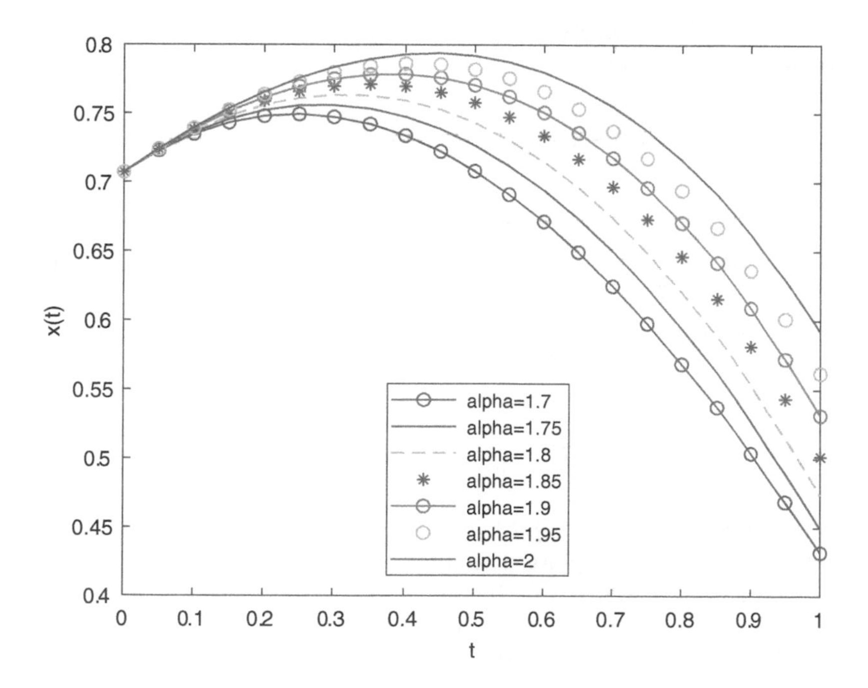

FIGURE 3.1
Behaviour of approximate solution for different values of $\alpha = 1.7, 1.75, 1.8, 1.85, 1.9, 1.95$ and 1 for case 1.

TABLE 3.1
Comparison between Results by Our Proposed Method and Exact Solution at $\alpha = 2$ and $n = 6$ for Case 1

t	Exact Solution	Present Method	Absolute Error
0.00	0.7071067	0.7071067	0
0.01	0.7106334	0.7106155	1.7830e5
0.02	0.7141419	0.7140699	7.1928e5
0.03	0.7176318	0.7174686	1.6320e4
0.04	0.7211028	0.7208102	2.9257e4
0.05	0.7245544	0.7240935	4.6092e4
0.06	0.7279862	0.7273171	6.6917e4
0.07	0.7313979	0.7304797	9.1822e4
0.08	0.7347890	0.7335800	1.2089e3
0.09	0.7381591	0.7366167	1.5423e3
0.1	0.7415079	0.7395886	1.9192e3

solutions from our proposed method, exact solution and absolute errors of proposed method.

From Table 3.1, it observed that results from the proposed method are accurate.

Case 2. In this case, we solve Lienard equation given by Eq. (3.3) with the following initial conditions [15,17]

$$x(0) = \varsigma = \sqrt{\frac{\theta}{2+\lambda}} \text{ and } x'(0) = \tau = 0, \tag{3.15}$$

where

$$\theta = 4\sqrt{\frac{3p^2}{(3q^2 - 16pr)}} \text{ and } \lambda = -1 + \frac{\sqrt{3}q}{\sqrt{(3q^2 - 16pr)}}. \tag{3.16}$$

For this case, we take $p = -1, q = 4$ and $r = 3$. The exact solution for integer order of Eq. (3.3) with initial condition in Eq. (3.15) is given as

$$x(t) = \sqrt{\frac{\theta \operatorname{sech}^2 \sqrt{-p}t}{2 + \lambda \operatorname{sech}^2 \sqrt{-p}t}}, \tag{3.17}$$

where θ and λ are given in Eq. (3.16).

In Figure 3.2, we have plotted an approximate solution for different values of $\alpha = 1.7, 1.75, 1.8, 1.85, 1.9, 1.95$ and 2.

From Figure 3.2, it is observed that the solution varies continuously from fractional to integer order. Periodic property is more important to FLE when α is close as 2. In Figure 3.3, we have shown the behaviour of approximate solution for close values of $\alpha = 1.95, 1.97, 1.99$ and 2.

From Figure 3.3, it is observed that the solution varies continuously from fractional order to integer order, and periodic behaviour of solution is seen.

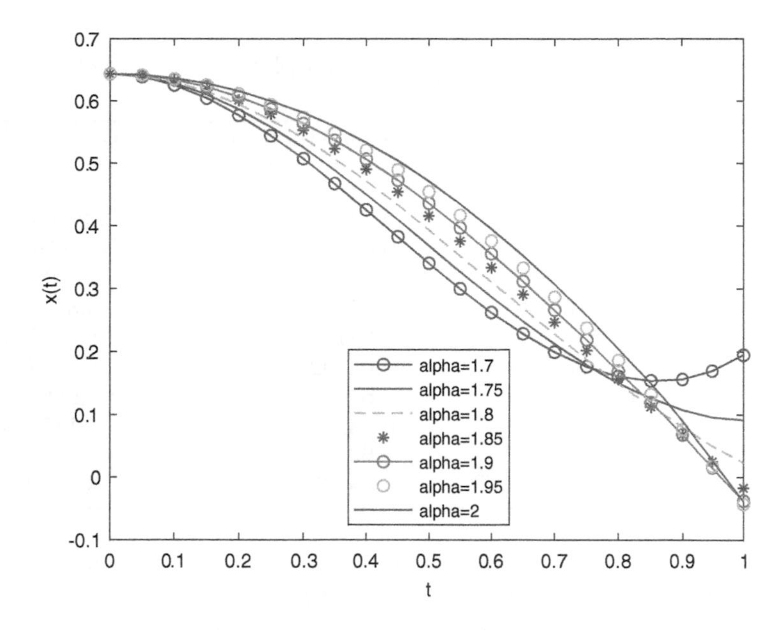

FIGURE 3.2
Behaviour of approximate solution for different values of $\alpha = 1.7, 1.75, 1.8, 1.85, 1.9, 1.95$ and 1 for case 2.

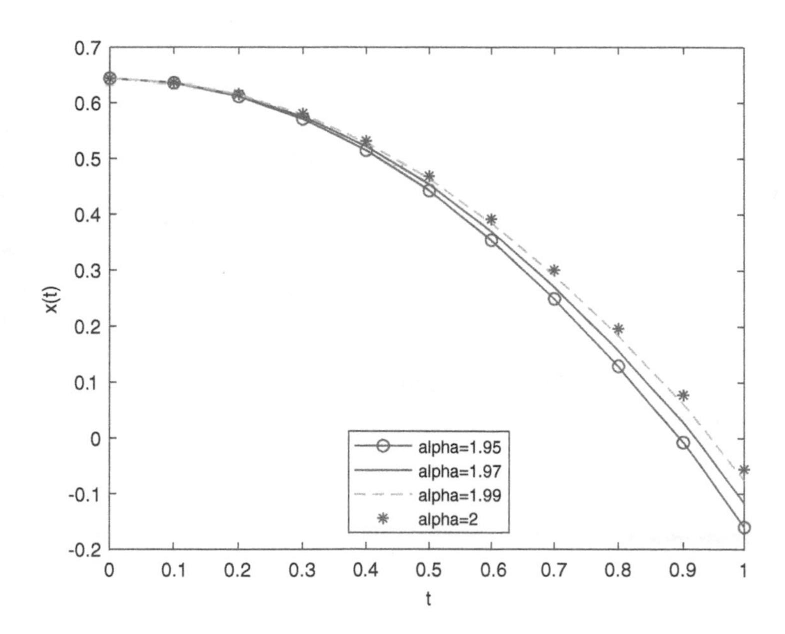

FIGURE 3.3
Behaviour of approximate solution for different values of $\alpha = 1.95, 1.97, 1.99$ and 2 for case 2.

TABLE 3.2
Comparison between Results by Our Proposed Method and Exact Solution at $\alpha = 2$ and $n = 6$ for Case 2

t	Exact Solution	Present Method	Absolute Error
0.00	0.6435942	0.6435942	0
0.01	0.6435565	0.6435241	3.2442e−5
0.02	0.6434434	0.6433126	1.3078e−4
0.03	0.6432551	0.6429586	2.9649e−4
0.04	0.6429915	0.6424606	5.3095e4
0.05	0.6426530	0.6418175	8.3545e4
0.06	0.6422396	0.6410284	1.2112e3
0.07	0.6417518	0.6400923	1.6594e3
0.08	0.6411897	0.6390086	2.1811e3
0.09	0.6405539	0.6377764	2.7774e3
0.1	0.6398446	0.6363954	3.4492e3

In Table 3.2, we have listed approximate solutions from our proposed method, exact solution and absolute errors of proposed method.

From Table 3.2, it is observed that results from the proposed method are accurate.

3.5 Conclusions

In this article, we have successfully applied our proposed method to analyze the FLE. Proposed method is good for computational purpose. From Tables 3.1 and 3.2, it is clear that our proposed method is accurate. Behaviours of approximate solution for different fractional order are shown through the figures, and it is observed that it converges to integer order solution. Proposed computational method can be used to solve various non-linear models in engineering and science. The fractional order solution of the Lienard equation will be very helpful to researchers working on the Lienard equation and fractional calculus. We can use different orthogonal polynomials operational matrix to achieve more accuracy.

3.6 Application

If we take $r = 0$, in FLE, then it reduces to fractional duffing equation. The special case of Lienard equation known as duffing equation is given as

$$x^{\alpha}(t) + 0.5y' + 25y + 25y^3 = 0, \quad 1 < \alpha \leq 2, \tag{3.18}$$

with initial conditions:

$$y(0) = 0.1, \quad y'(0) = 0. \tag{3.19}$$

We can use a similar algorithm for fractional duffing equation as for FLE.

Let us take the following approximation

$$x(t) = \sum_{i=0}^{n} c_i \psi_i(t) = C^T \psi(t).$$

Using this approximation in Eq. (3.18), we get

$$C^T D^{(\alpha)} \psi(t) + 0.5 C^T D^{(1)} \psi(t) \quad + \quad + \; 25 C^T \psi(t) + 25 \left(C^T \psi(t)\right)^3 = 0. \tag{3.20}$$

The residual for Eq. (3.20) can be written as

$$R_n(t) = C^T D^{(\alpha)} \psi(t) + 0.5 C^T D^{(1)} \psi(t) \quad + \quad + \; 25 C^T \psi(t) + 25 \left(C^T \psi(t)\right)^3, \tag{3.21}$$

where $D^{(\alpha)}$ and $D^{(1)}$ are operational matrices of differentiations of order α and 1, respectively, given by Eq. (3.8). Now, collocate Eq. (3.21) at $n-1$ points, given by $x_i = \frac{i}{n}$, $i = 0, 1, 2, \ldots, n-2$.

Initial conditions given in Eq. (3.19) are given by

$$C^T \psi(0) = 0.1, \qquad C^T D^{(1)} \psi(0) = 0. \tag{3.22}$$

Using collocation points in Eq. (3.21) along with the Eq. (3.22), we get a system of $n+1$ non-linear algebraic equations whose solution gives the unknown in the approximation. Using these values of unknowns, we get the approximate solution for fractional duffing equation.

Appendix

The operational matrix of differentiations for Legendre scaling functions at $\alpha = 1$ and $n = 6$ is given as

$$D^{(1)} = \begin{bmatrix} 0 & 0 & 0 & 0 & 0 & 0 & 0 \\ 3.4641 & 0 & 0 & 0 & 0 & 0 & 0 \\ 0 & 7.7460 & 0 & 0 & 0 & 0 & 0 \\ 5.2915 & 0 & 11.8322 & 0 & 0 & 0 & 0 \\ 0 & 10.3923 & 0 & 15.8745 & 0 & 0 & 0 \\ 6.6332 & 0 & 14.8324 & 0 & 19.8997 & 0 & 0 \\ 0 & 12.4900 & 0 & 19.0788 & 0 & 23.9165 & 0 \end{bmatrix}$$

The operational matrix of differentiations for Legendre scaling functions at $\alpha = 1.9$ and $n = 6$ is given as

$$D^{(1.9)} = \begin{bmatrix} 0 & 0 & 0 & 0 & 0 & 0 & 0 \\ 0 & 0 & 0 & 0 & 0 & 0 & 0 \\ 25.6409 & 2.1148 & -0.7926 & 0.4346 & -0.2802 & 0.1981 & -0.1486 \\ -7.2235 & 76.2796 & 7.4852 & -3.1550 & 1.8640 & -1.2649 & 0.9279 \\ 109.1210 & -5.8823 & 144.6079 & 16.9250 & -7.7855 & 4.8572 & -3.4232 \\ -30.4711 & 256.8657 & 1.0768 & 229.6718 & 31.1655 & -15.3365 & 9.9770 \\ 267.2947 & -33.2173 & 432.3633 & 14.1084 & 330.2464 & 50.8829 & -26.4286 \end{bmatrix}$$

References

[1] Liénard, A. (1928). Etude des oscillations entretenues. *Rev. Gen. Electr.*, 23, 901–912.

[2] Guckenheimer, J. (1980). Dynamics of the van der pol equation. *IEEE Trans. Circ. Syst.*, 27, 938–989.

[3] Zhang, Z. F., Ding, T. & Huang, H. W. (1985). *Qualitative Theory of Differential Equations.* Science Press, Peking.

[4] Hale, J. K. (1980). *Ordinary Differential Equations.* Wiley, New York.

[5] Kong, D. (1995). Explicit exact solutions for the Lienard equation and its applications. *Phys. Lett. A*, 196, 301–306.

[6] Singh, H. (2017). A new stable algorithm for fractional Navier-Stokes equation in polar coordinate. *Int. J. Appl. Comp. Math.*, 3, 3705–3722.

[7] Panda, R. & Dash, M. (2006). Fractional generalized splines and signal processing. *Signal Process.*, 86, 2340–2350.

[8] Singh, H. (2017). Operational matrix approach for approximate solution of fractional model of Bloch equation. *J. King Saud Univ. Sci.*, 29, 235–240.

[9] Bagley, R. L. & Torvik, P. J. (1983). A theoretical basis for the application of fractional calculus to viscoelasticity. *J. Rheol.*, 27, 201–210.

[10] Bagley, R. L. & Torvik, P. J. (1983). Fractional calculus a differential approach to the analysis of viscoelasticity damped structures. *AIAA J.*, 21 (5), 741–748.

[11] Bagley, R. L. & Torvik, P. J. (1985). Fractional calculus in the transient analysis of viscoelasticity damped structures. *AIAA J.*, 23, 918–925.

[12] Magin, R. L. (2004). Fractional calculus in bioengineering. *Crit. Rev. Biomed. Eng.*, 32, 1–104.

[13] Robinson, A. D. (1981). The use of control systems analysis in neurophysiology of eye movements. *Ann. Rev. Neurosci.*, 4, 462–503.

[14] Feng, Z. (2002). On explicit exact solutions for the Lienard equation and its applications. *Phys. Lett. A*, 239, 50–56.

[15] Matinfar, M., Hosseinzadeh, H. & Ghanbari, M. (2008). A numerical implementation of the variational iteration method for the Lienard equation. *World J. Model. Simul.*, 4, 205–210.

[16] Matinfar, M., Mahdavi, M. & Raeisy, Z. (2011). Exact and numerical solution of Lienard's equation by the variational homotopy perturbation method. *J. Inf. Comput. Sci.*, 6 (1), 73–80.

[17] Kumar, D., Agarwal, R. P. & Singh, J. (2017). A modified numerical scheme and convergence analysis for fractional model of Lienard's equation. *J. Comput. Appl. Math.*, 339, 405–413.

[18] Tohidi, E., Bhrawy, A. H. & Erfani, K. (2013). A collocation method based on Bernoulli operational matrix for numerical solution of generalized pantograph equation. *Appl. Math. Model.*, 37, 4283–4294.

[19] Singh, H., Srivastava, H. M. & Kumar, D. (2017). A reliable numerical algorithm for the fractional vibration equation. *Chaos, Solitons Fractals*, 103, 131–138.

[20] Kazem, S., Abbasbandy, S. & Kumar, S. (2013). Fractional order Legendre functions for solving fractional-order differential equations. *Appl. Math. Model.*, 37, 5498–5510.

[21] Singh, H. (2018). An efficient computational method for the approximate solution of nonlinear Lane-Emden type equations arising in astrophysics. *Astrophys. Space Sci.*, 363 (4), 363–371.

[22] Lakestani, M., Dehghan, M. & Pakchin, S. I. (2012). The construction of operational matrix of fractional derivatives using B-spline functions. *Commun. Nonlinear Sci. Numer. Simul.*, 17, 1149–1162.

[23] Wu, J. L. (2009). A wavelet operational method for solving fractional partial differential equations numerically. *Appl. Math. Comput.*, 214, 31–40.

[24] Yousefi, S. A., Behroozifar, M. & Dehghan, M. (2011). The operational matrices of Bernstein polynomials for solving the parabolic equation subject to the specification of the mass. *J. Comput. Appl. Math.*, 235, 5272–5283.

[25] Singh, H. (2018). Approximate solution of fractional vibration equation using Jacobi polynomials. *Appl. Math. Comput.*, 317, 85–100.

[26] Singh, C. S., Singh, H., Singh, V. K. & Singh, Om P. (2016). Fractional order operational matrix methods for fractional singular integro-differential equation. *Appl. Math. Model.*, 40, 10705–10718.

[27] Singh, H. (2016). A new numerical algorithm for fractional model of Bloch equation in nuclear magnetic resonance. *Alex. Eng. J.*, 55, 2863–2869.

4

A New Approximation Scheme for Solving Ordinary Differential Equation with Gomez–Atangana–Caputo Fractional Derivative

Toufik Mekkaoui and Zakia Hammouch

Faculty of Sciences and Techniques, Moulay Ismail University of Meknes

Devendra Kumar

University of Rajasthan

Jagdev Singh

JECRC University

CONTENTS

4.1 Introduction

Fractional calculus (FC) is a generalization of classical calculus. FC is an excellent mathematical tool that describes the systems memory and their hereditary properties. Theory and applications of FC have become extremely useful and important in different science fields, such as in biological processes modelling, applied mathematics, physics, and engineering, because of their memory property [1–12]. There are different definitions of fractional derivatives, like

Riemann–Liouville, Grunwald–Letnikov, Caputo, Weyl, Hadamard, Erdelyi–Kober, Riesz, Coimbra, Jumarie, Fabrizio Caputo, Atangana–Baleanu, among others [13–24]. Recently, Gomez and Atangana presented a novel version of fractional derivatives with two orders: power law and exponential decay. In this chapter, we will present a novel numerical method for solving ordinary differential equations with fractional order in Gomez–Atangana–Caputo sense. We shall present the application of the new method to some well-known problems, including an equation having exact solution, a Lotka–Volterra model and a fractional-order Chen system.

4.2 A New Numerical Approximation

First, we present the basic definitions of some derivatives with fractional order, namely Caputo, Fabrizio–Caputo and Gomez–Atangana–Caputo, which combines power law and exponential decay.

Definition 4.1 The Caputo fractional derivatives D^{α} of a function $y(t)$ of any real number α such that $0 < \alpha \leq 1$, for $t > 0$ is defined as

$$D^{\alpha}y(t) = \frac{1}{\Gamma(1-\alpha)} \int_0^t y'(\tau)(t-\tau)^{-\alpha} d\tau, \tag{4.1}$$

Definition 4.2 Let y be in $H^1(a, b)$, Then the fractional Caputo–Fabrizio derivative, with $0 \leq \alpha \leq 1$, and $a \in [-\infty, t]$ is given as follows:

$${}_0^{CF}D_t^{\alpha}y(t) = \frac{M(\alpha)}{1-\alpha} \int_a^t y'(\tau) \exp[-\alpha \frac{t-\tau}{1-\alpha}] d\tau. \tag{4.2}$$

Definition 4.3 The Gomez–Atangana fractional derivative with bi-order (α, β) in the Caputo sense (GAC) is given as in [1]

$${}_0^{GAC}D_t^{\alpha,\beta}y(t) = \frac{M(\alpha)}{(1-\alpha)} \frac{1}{\Gamma(1-\beta)} \int_0^t y'(\tau) \exp\left(-\frac{\alpha}{1-\alpha}(t-\tau)^{-\alpha}\right)(t-\tau)^{-\beta} d\tau. \tag{4.3}$$

where $0 < \alpha,\ \beta \leq 1$.

Now, we describe the new numerical approximation for solving ordinary differential equations with GAC fractional derivatives. Consider the general non-linear fractional differential equation

$$\begin{cases} {}_0^{GAC}D_t^{\alpha,\beta}y(t) = f(t, y(t)) \qquad t \in [0, T], \\ y(0) = y_0, \end{cases} \tag{4.4}$$

where $0 < \alpha,\ \beta \leq 1$ and $\alpha + \beta = 1$.

We take a uniform mesh on the interval $[0, T]$ and the nodes $0, 1, ..., N$, where N is an arbitrary positive integer and the time step size $h = \frac{T}{N}$, $t_{k+1} = t_k + h$ and denoting y_k the numerical approximation of $y(t_k)$.

Using (4.1) at $t = t_{n+1}$, we get

$$
\begin{aligned}
{}_{0}^{GAC}D_t^{\alpha,1-\alpha}y(t_{n+1}) &= \frac{M(\alpha)}{1-\alpha}\frac{1}{\Gamma(\alpha)}\int_0^{t_{n+1}} y'(\tau)\exp\left(-\frac{\alpha}{1-\alpha}(t_{n+1}-\tau)^{\alpha}\right) \\
&\quad \times (t_{n+1}-\tau)^{\alpha-1}d\tau \qquad (4.5) \\
&= \frac{M(\alpha)}{1-\alpha}\frac{1}{\Gamma(\alpha)}\sum_{k=0}^{n}\int_{t_k}^{t_{k+1}} y'(\tau)\exp\left(-\frac{\alpha}{1-\alpha}(t_{n+1}-\tau)^{\alpha}\right) \\
&\quad \times (t_{n+1}-\tau)^{\alpha-1}d\tau. \qquad (4.6)
\end{aligned}
$$

By approximating the derivative $y'(\tau)$ on each interval $[t_k, t_{k+1}]$ by

$$
\frac{y(t_{k+1}) - y(t_k)}{t_{k+1} - t_k} = \frac{y(t_{k+1}) - y(t_k)}{h} := \Delta(y_{k+1}), \qquad (4.7)
$$

we get

$$
\begin{aligned}
{}_{0}^{GAC}D_t^{\alpha,1-\alpha}y_{n+1}(t) &= \frac{M(\alpha)}{1-\alpha}\frac{1}{\Gamma(\alpha)}\Delta(y_{k+1})\int_0^{t_{n+1}} \exp\left(-\frac{\alpha}{\alpha}(t_{n+1}-\tau)^{\alpha}\right) \\
&\quad \times (t_{n+1}-\tau)^{\alpha-1}d\tau. \qquad (4.8)
\end{aligned}
$$

where

$$
\begin{aligned}
&\int_0^{t_{n+1}} \exp\left(-\frac{\alpha}{1-\alpha}(t_{n+1}-\tau)^{\alpha}\right)(t_{n+1}-\tau)^{\alpha-1}d\tau \\
&= \frac{1-\alpha}{\alpha^2}\exp\left(-\frac{\alpha}{1-\alpha}(t_{n+1}-t_{k+1})^{\alpha}\right) \\
&\quad - \frac{1-\alpha}{\alpha^2}\exp\left(-\frac{\alpha}{1-\alpha}(t_{n+1}-t_k)^{\alpha}\right) \\
&= \frac{1-\alpha}{\alpha^2}\left[\exp\left(-\frac{\alpha}{1-\alpha}h^{\alpha}(n-k)^{\alpha}\right)\right. \\
&\quad \left. - \exp\left(-\frac{\alpha}{1-\alpha}h^{\alpha}(n+1-k)^{\alpha}\right)\right], \qquad (4.9)
\end{aligned}
$$

Replacing equation (4.9) into equation (4.8), we finally obtain the following:

$$
\begin{aligned}
{}_{0}^{GAC}D_t^{\alpha,1-\alpha}y_{n+1}(t) &= \frac{M(\alpha)}{\alpha^2\Gamma(\alpha)}\sum_{k=0}^{n-1}\Delta(y_{k+1})\left(\exp\left(-\frac{\alpha}{1-\alpha}h^{\alpha}(n-k)^{\alpha}\right)\right. \\
&\quad \left. - \exp\left(-\frac{\alpha}{1-\alpha}h^{\alpha}(n+1-k)^{\alpha}\right)\right) \\
&\quad + \Delta(y_{n+1})\left(1 - \exp\left(-\frac{\alpha}{1-\alpha}\right)\right). \qquad (4.10)
\end{aligned}
$$

4.2.1 Error Estimate

Now, we analyze the error estimate of the suggested approximation method. This will be established in the following theorem.

Theorem 4.1 *Let y be a function in the space $C^{n+1}([0,T])$ and $\alpha \in (0,1)$. Then we have*

$$ {}_{0}^{GAC}D_t^{\alpha,1-\alpha}y(t) = \frac{M(\alpha)}{\alpha^2\Gamma(\alpha)}\sum_{k=0}^{n-1}\Delta(y_{k+1})H_{n,k} + \Theta(h^{\alpha+1}). \tag{4.11} $$

where

$$ H_{n,k} = \left(\exp\left(-\frac{\alpha}{1-\alpha}h^\alpha(n-1-k)^\alpha\right) - \exp\left(-\frac{\alpha}{1-\alpha}h^\alpha(n-k)^\alpha\right)\right), \tag{4.12} $$

Proof 1 *From the above, we have*

$$ {}_{0}^{GAC}D_t^{\alpha,1-\alpha}y(t) = \frac{M(\alpha)}{\alpha^2\Gamma(\alpha)}\sum_{k=0}^{n-1}\left(\Delta(y_{k+1})H_{n,k} + \Theta(h^{\alpha+1})\right), \tag{4.13} $$

On the other hand, we have

$$ \begin{aligned} &\sum_{k=0}^{n-1}\left(\exp\left(-\frac{\alpha}{1-\alpha}h^\alpha(n-1-k)^\alpha\right) - \exp\left(-\frac{\alpha}{1-\alpha}h^\alpha(n-k)^\alpha\right)\right) \\ &\qquad = 1 - \exp\left(-\frac{-\alpha}{1-\alpha}h^\alpha n^\alpha\right). \end{aligned} \tag{4.14} $$

Then, we get

$$ \begin{aligned} {}_{0}^{GAC}D_t^{\alpha,1-\alpha}y(t) = {}& \frac{M(\alpha)}{\alpha^2\Gamma(\alpha)}\sum_{k=0}^{n-1}\Delta(y_{k+1})H_{n,k} + \frac{M(\alpha)}{\alpha^2\Gamma(\alpha)}\Theta(h) \\ & \times\left(1 - \exp\left(-\frac{\alpha}{1-\alpha}h^\alpha n^\alpha\right)\right). \end{aligned} \tag{4.15} $$

Moreover, we have for h small and $\alpha \in (0,1)$:

$$ \exp\left(-\frac{\alpha}{1-\alpha}h^\alpha n^\alpha\right) \simeq 1 - \frac{\alpha}{1-\alpha}h^\alpha n^\alpha. \tag{4.16} $$

Consequently, we obtain

$$ {}_{0}^{GAC}D_t^{\alpha,1-\alpha}y(t) = \frac{M(\alpha)}{\alpha^2\Gamma(\alpha)}\sum_{k=0}^{n-1}\Delta(y_{k+1})H_{n,k} + \frac{M(\alpha)}{(1-\alpha)\Gamma(1+\alpha)}n^\alpha h^{1+\alpha}. \tag{4.17} $$

setting $\Theta(h^{\alpha+1}) = \frac{M(\alpha)}{(1-\alpha)\Gamma(1+\alpha)}n^\alpha h^{1+\alpha}$, the proof is completed.

4.3 Application

In this section, we illustrate the new approximation by solving some fractional differential equations.

4.3.1 Example 1

Let us consider the following fractional ordinary differential equation

$$\begin{cases} {}_{0}^{GAC}D_t^{0.5,0.5}y(t) = -\frac{8}{3}\exp\left(t+\frac{1}{4}\right)\left(\operatorname{erf}\left(\frac{1}{2}\right) - \operatorname{erf}\left(\sqrt{t}+\frac{1}{2}\right)\right), \\ y(0) = 1. \end{cases} \tag{4.18}$$

For which the exact solution is obtained as $y(t) = \exp(t)$. Using the proposed approximation described earlier with $h = 0.01$, we obtain a numerical solution presented in Figure 4.1 versus the exact solution. The comparison shows an excellent agreement between the exact and numerical solutions even for a large value of time t.

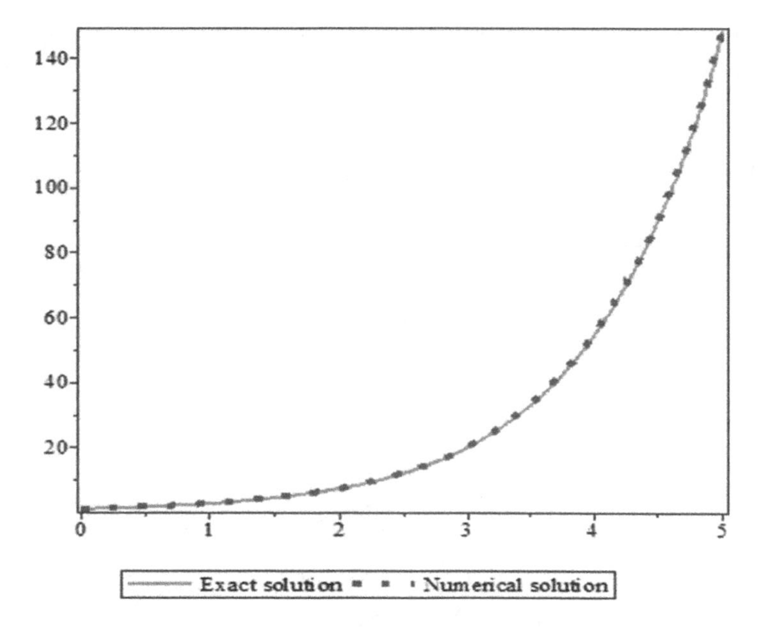

FIGURE 4.1
Profile of exact solution vs numerical solution of (4.18).

4.3.2 Example 2

We consider the fractional-ordered Lotka–Volterra equations, also known as predator–prey equations, frequently used to describe the dynamics of biological systems in which two species interact, one as a predator and the other as a prey. The populations change through time according to the pair of equations

$$\begin{cases} {}_{0}^{GAC}D_{t}^{0.97,0.03}x(t) = ax(t) - bx(t)y(t), \\ {}_{0}^{GAC}D_{t}^{0.97,0.03}y(t) = dx(t)y(t) - cy(t), \\ x(0) = 15, \qquad y(0) = 8. \end{cases} \tag{4.19}$$

In order to check the effect of combining power law and exponential decay into the definition of fractional derivative, we will present graphically the numerical solutions for the Lotka–Volterra system for GAC, Caputo Fabrizio and Caputo derivatives, respectively, for $\alpha = 0.97$, $a = 0.95$, $b = 0.25$, $c = 2.45$, $d = 0.25$ and $h = 0.01$. We can observe from Figure 4.2 that the plots are almost similar.

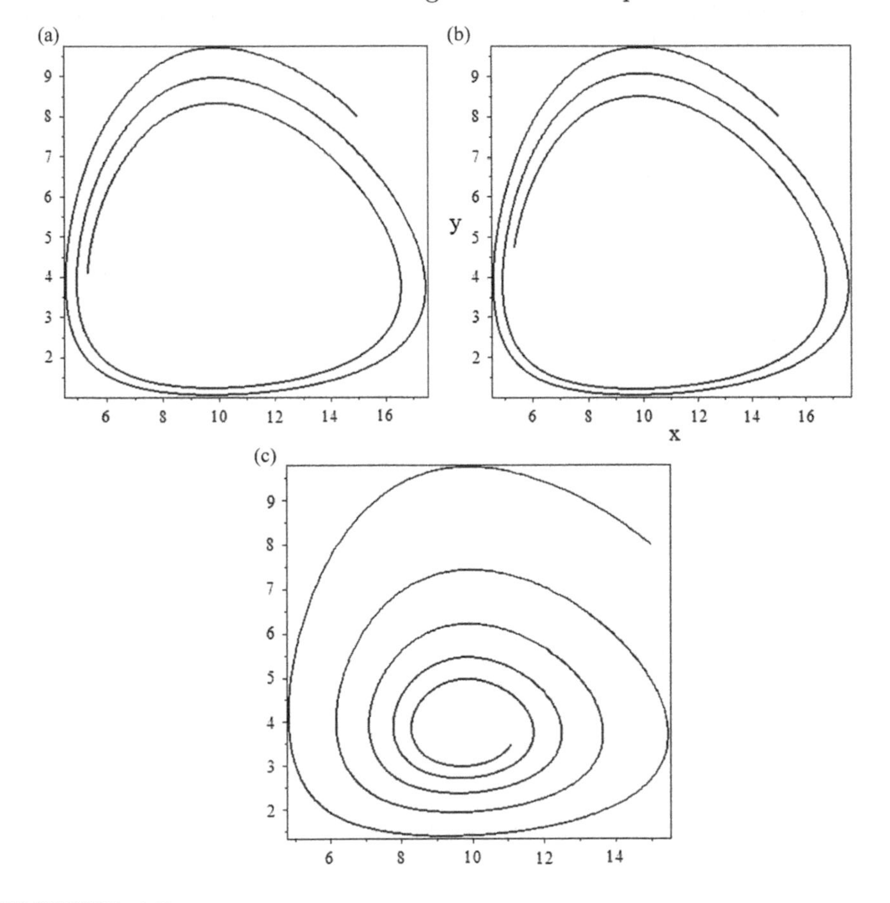

FIGURE 4.2
Phase planes profiles x-y for system (4.19): (a) GAC derivative, (b) Caputo Fabrizio derivative and (c) Caputo derivative.

4.3.3 Example 3

In this example, we solve the well-known Chen chaotic system, where the fractional derivative is that of Gomez–Atangana in Caputo sense. Chen dynamical system [25] is described by the following system of differential equations

$$\begin{cases} {}_0^{GAC}D_t^{0.95,0.05}x(t) = a(y(t) - x(t)), \\ {}_0^{GAC}D_t^{0.95,0.05}y(t) = (c-a)x(t) - x(t)z(t) + cy(t), \\ {}_0^{GAC}D_t^{0.95,0.05}z(t) = x(t)y(t) - bz(t), \\ x(0) = -0.1, \qquad y(0) = 0.5, \qquad z(0) = -0.6. \end{cases} \tag{4.20}$$

The simulation results demonstrate that chaos indeed exist in the fractional order Chen system with $\alpha = 0.95$ and $a = 40$, $b = 3$ and $c = 28$. We can see that the chaotic attractors and the phase planes within GAC derivatives are similar and look like those obtained with Caputo–Fabrizio and Caputo fractional derivatives (see Figures 4.3–4.6).

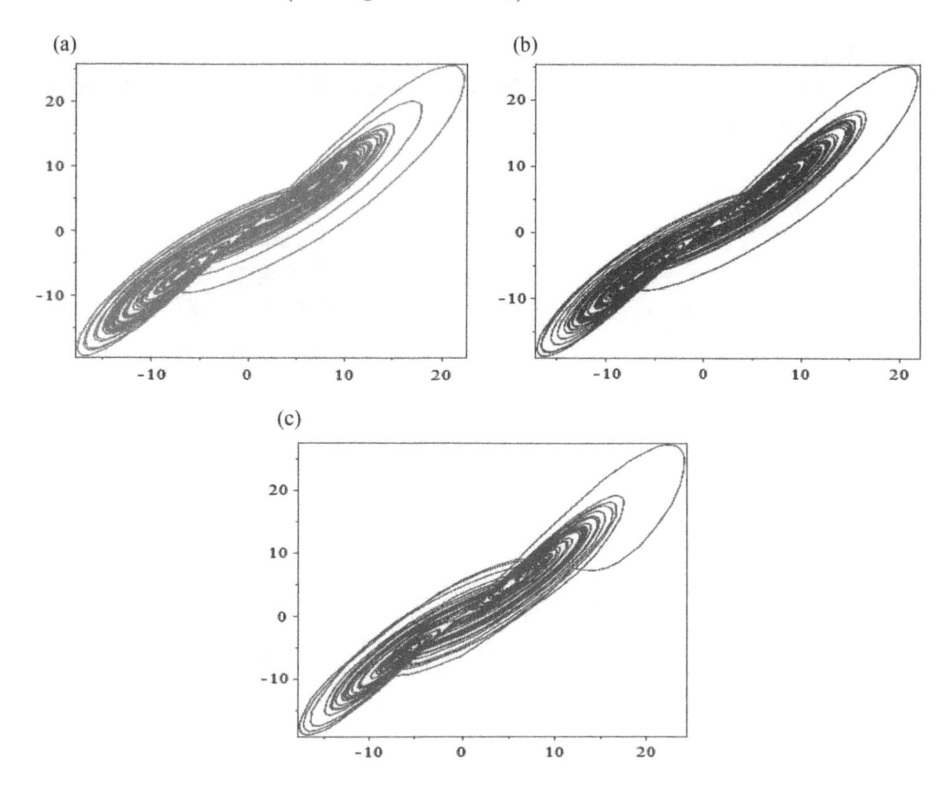

FIGURE 4.3
Phase plan x-y for system (4.20) : (a) GAC derivative, (b) Caputo-Fabrizio derivative and (c) Caputo derivative.

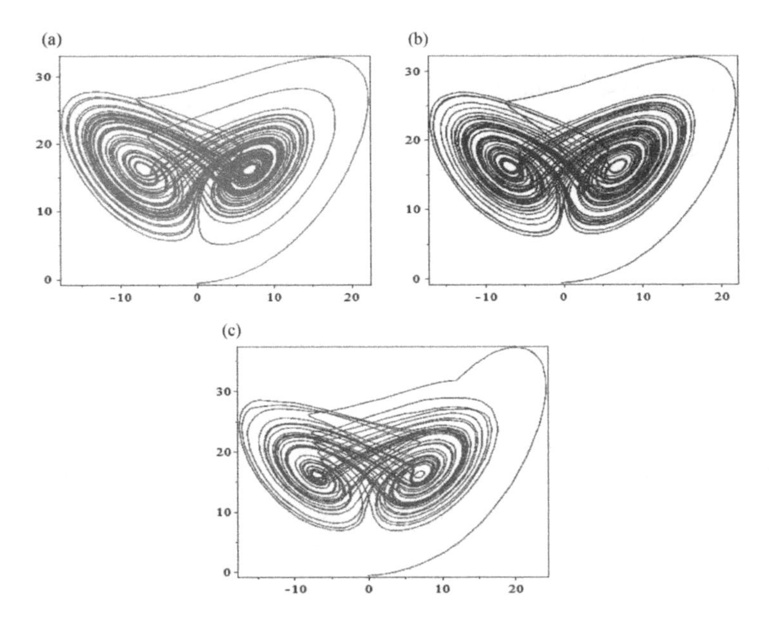

FIGURE 4.4
Phase plan x-z for system (4.20) : (a) GAC derivative, (b) Caputo-Fabrizio derivative and (c) Caputo derivative.

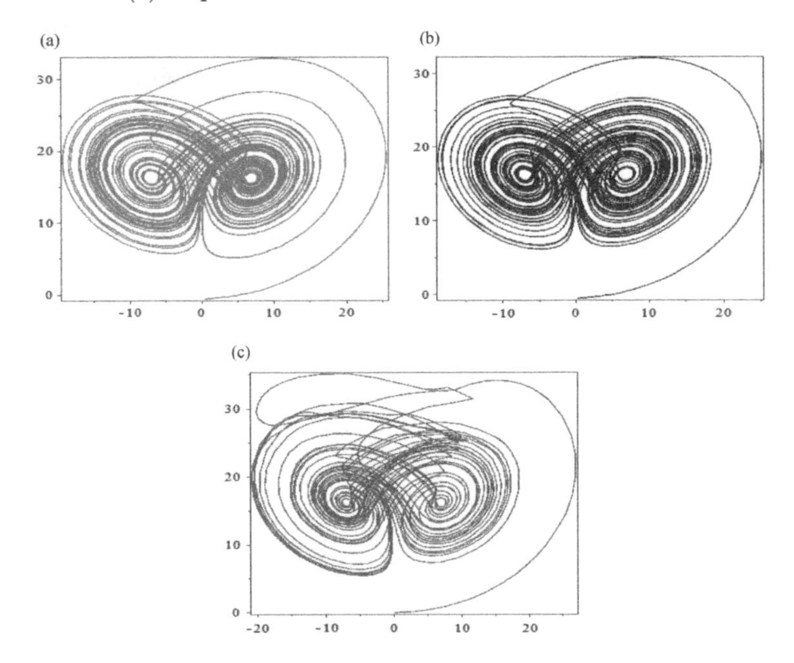

FIGURE 4.5
Phase plan y-z for system (4.20) : (a) GAC derivative, (b) Caputo-Fabrizio derivative and (c) Caputo derivative.

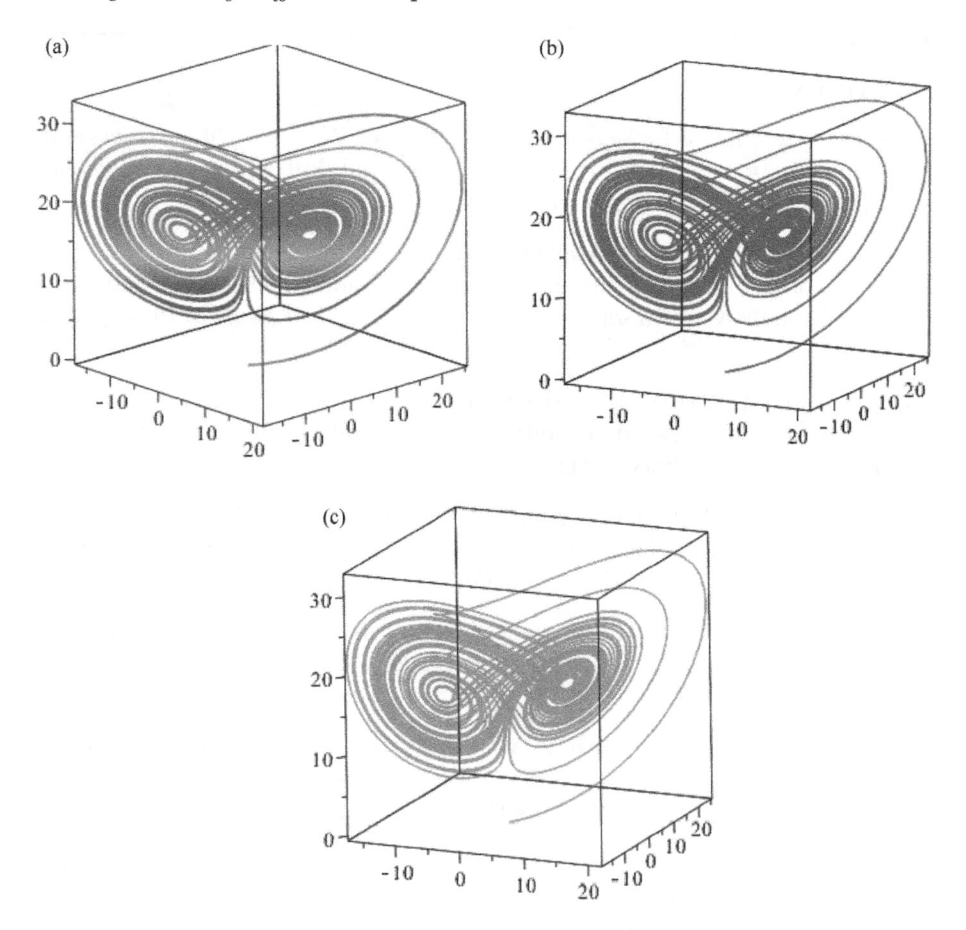

FIGURE 4.6
Attractor x-y-z for system (4.20) : (a) GAC derivative, (b) Caputo-Fabrizio derivative and (c) Caputo derivative.

4.4 Conclusion

Within the framework of the fractional differentiation with combined power law and exponential decay, a new numerical approximation was suggested to solve non-linear fractional differential equations. The numerical scheme is introduced and the error estimate is given. The method is highly accurate and efficient. Some examples are given to prove the efficiency and accuracy of the numerical method.

References

[1] Gomez-Aguilar, J.F. & Atangana, A. (2017). New insight in fractional differentiation: Power, exponential decay and Mittag-Leffler laws and applications. *The European Physical Journal Plus*, (132), 13.

[2] Gomez-Aguilar, J.F. & Atangana, A. (2017). Fractional Hunter-Saxton equation involving partial operators with bi-order in Riemann-Liouville and Liouville-Caputo sense. *The European Physical Journal Plus*, (132), 100.

[3] Atangana, A. (2016). Derivative with two fractional orders: A new avenue of investigation toward revolution in fractional calculus. *The European Physical Journal Plus*, (131), 373.

[4] Atangana, A. & Gomez-Aguilar, J.F. (2017). Hyperchaotic behaviour obtained via a nonlocal operator with exponential decay and Mittag-Leffler laws. *Chaos, Solitons and Fractals*, (102), 285–294.

[5] Ucar, S. et al. (2019). Mathematical analysis and numerical simulation for a smoking model with Atangana–Baleanu derivative. *Chaos, Solitons and Fractals*, (118), 300–306.

[6] Owolabi, M.K. & Atangana, A. (2019). High-order solvers for space-fractional differential equations with Riesz derivative. *Discrete and Continuous Dynamical Systems-S*, (12), 567–590.

[7] Singh, J. et al. (2018). An efficient computational approach for time-fractional Rosenau–Hyman equation. *Neural Computing and Applications*, (30), 3063–3070.

[8] Kumar, D. et al. Analysis of a fractional model of the Ambartsumian equation. *The European Physical Journal Plus*, (133), 259.

[9] Kumar, D., Singh, J. & Baleanu, D. (2018). A new numerical algorithm for fractional Fitzhugh–Nagumo equation arising in transmission of nerve impulses. *Nonlinear Dynamics*, (91), 307–317.

[10] Srivastava, H.M., Kumar, D. & Singh, J. (2017). An efficient analytical technique for fractional model of vibration equation. *Applied Mathematical Modelling*, (45), 192–204.

[11] Kumar, D., Singh, J. & Baleanu, D. (2017). A fractional model of convective radial fins with temperature-dependent thermal conductivity. *Romanian Reports in Physics*, (69), 103.

[12] Singh, J., Kumar, D. & Nieto, J.J. (2017). Analysis of an El Nino-Southern Oscillation model with a new fractional derivative. *Chaos, Solitons and Fractals*, (99), 109–115.

[13] Jarad, F., Abdeljawad, T. & Hammouch, Z. (2018). On a class of ordinary differential equations in the frame of Atangana–Baleanu fractional derivative. *Chaos, Solitons and Fractals*, (117), 16–20.

[14] Baskonus, H.M., Mekkaoui, T., Hammouch, Z. & Bulut, H. (2015). Active control of a chaotic fractional order economic system. *Entropy*, (17), 5771–5783.

[15] Baskonus, H.M., Hammouch, Z., Mekkaoui, T. & Bulut, H. Chaos in the fractional order logistic delay system: Circuit realization and synchronization. *AIP Conference Proceedings*, 2016, 1.

[16] Singh, J., Kumar, D., Hammouch, Z. & Atangana, A. (2018). A fractional epidemiological model for computer viruses pertaining to a new fractional derivative. *Applied Mathematics and Computation*, (316), 504–515.

[17] Hammouch, Z. & Mekkaoui, T. (2014). Chaos synchronization of a fractional nonautonomous system. *Nonautonomous Dynamical Systems*, (1), (2014).

[18] Hammouch, Z. & Mekkaoui, T. (2015). Control of a new chaotic fractional-order system using Mittag-Leffler stability. *Nonlinear Studies*, 22(4), 565–577.

[19] Hammouch, Z., Mekkaoui, T. & Belgacem, F. (2014). Numerical simulations for a variable order fractional Schnakenberg model. *AIP Conference Proceedings*, 2014, 1637.

[20] Hammouch, Z. & Mekkaoui, T. (2018). Circuit design and simulation for the fractional-order chaotic behavior in a new dynamical system. *Complex and Intelligent Systems*, (1), 1–10.

[21] Asif, N.A., Hammouch, Z., Riaz, M.B. & Bulut, H. (2018). Analytical solution of a Maxwell fluid with slip effects in view of the Caputo-Fabrizio derivative. *The European Physical Journal Plus*, 133(7), 272.

[22] Hammouch, Z., Mekkaoui, T. & Agarwal, P. (2018). Optical solitons for the Calogero-Bogoyavlenskii-Schiff equation in (2+1) dimensions with time-fractional conformable derivative. *The European Physical Journal Plus*, 133(7), 248.

[23] Chand, M., Hammouch, Z., Asamoah, J.K. & Baleanu, D. (2019). Certain fractional integrals and solutions of fractional kinetic equations involving the product of s-function. *Mathematical Methods in Engineering*, 213–244.

[24] Ait Touchent, K., Hammouch, Z., Mekkaoui, T. & Belgacem, F.B.M. (2018). Implementation and convergence analysis of homotopy perturbation coupled with sumudu transform to construct solutions of local-fractional PDEs. *Fractal and Fractional*, 2(3), 22.

[25] Chunguang, L. & Chen, G. (2004). Chaos in the fractional order Chen system and its control. *Chaos, Solitons and Fractals*, (22)3, 549–554.

5

Fractional Optimal Control of Diffusive Transport Acting on a Spherical Region

Derya Avci and Necati Ozdemir
Balikesir University

Mehmet Yavuz
Necmettin Erbakan University

CONTENTS

5.1 Introduction

Fractional calculus (FC) has a growing interest and application in the modelling of a variety of dynamical systems in recent years. It has been demonstrated that many real physical processes can be modelled more accurately by using fractional derivatives and integrals [1–7]. The main advantage of this fact is that fractional order models provide the hereditary and memory effects that naturally arise in physical materials and dynamical systems. These effects are common to many physical processes, yet they are simply neglected by the classical theory. Anomalous diffusion and heat conduction in fractal porous media, viscoelastic and thermoelastic structures, non-Brownian motions of particles, biomechanics and control of the dynamic processes are only some of the applications of FC.

Optimal control problems of dynamic systems, which is governed by fractional diffusion-wave equations, is one of the popular application topics of FC.

In this chapter, optimal control problem of an anomalous diffusion equation is defined by a time–space FDE.

The main objective of classical optimal control problems is that foundation of the admissible control functions that minimize the performance index functional (cost function) subject to the system dynamics described by the state and control variables. If an optimal control problem in which either the performance index or the differential equations governing the dynamics of the system or both contain at least one fractional derivative term, this problem is called as a fractional optimal control problem (FOCP) [8]. The formulation in [8] is known as the first formulation of FOCPs because the author proposed a general and clear formulation for a class of FOCPs and also obtained the fractional Euler–Lagrange equations, which describe the necessary optimality conditions, by using the calculus of variations, the Lagrange multiplier technique and the formula of fractional integration by parts. Main contribution of this chapter to literature is that the numerical formulation scheme and the resulting equations are not only similar to classical theory but also more general so that it has been a basis for a lot different type of new FOCP formulations. Also, [9] is basically same as the previous work. But the problem is formulated in terms of Caputo fractional derivative, which is generally preferred among engineers and scientists because of the initial conditions that have real physical interpretation. In addition, an iterative numerical scheme by discretization of time domain is used to find the numerical solutions of FOCP in this chapter. Moreover, the dynamic system and performance index are defined with the state and control functions with constant coefficient, i.e. time invariant. In the next study, Agrawal expanded this formulation for the variable coefficient case, i.e. time-varying and used Volterra integral equations to find the numerical solutions [10]. Agrawal considered the FOCP for a class of continuum systems [11]. By this study, the extension of the study in [9] to the vector case was aimed, and also a numerical scheme based on the discretization of only the space domain was intended.

The necessary optimality conditions obtained by Lagrange multiplier technique consist of both forward and backward fractional derivatives. Therefore, the numerical solutions of FOCPs are found by using discretization of both forward and backward operators. It is clear to see Grünwald–Letnikov fractional derivative is based on this consideration and so is widely used as a numerical scheme in the literature. Agrawal applied Grünwald–Letnikov definition for both of time-invariant and -varying types [12]. Furthermore, Baleanu et al. presented a modified Grünwald–Letnikov approximation for the time-varying type of this problem [13]. Bisvas and Sen studied on the formulation of FOCPs for which the values of state function at the final time were specified and a numerical solution method based on Grünwald-Letnikov (GL) definition was applied to solve the equations [14].

In FOCPs, eigenfunction expansion techniques are considered to eliminate the space variable and to express the equations in terms of time variables only. This leads to a set of decoupled equations that can be solved independently and easily. For a class of distributed systems, the formulation and numerical solutions of FOCP were obtained by using an eigenfunction expansion technique [15]. The state and control functions are represented by the eigenfunction series by this technique. Thus, the main FOCP reduces to multi-FOCPs in which each FOCP can be solved independently. Because of this advantage, eigenfunction expansion method is widely used for the formulation of FOCPs that are also defined in different coordinate systems. For example, the FOCP for a class of distributed systems formulated in polar and cylindrical coordinates for axial symmetric case was considered in [16–18]. In these papers, they defined system dynamics in terms of Riemann–Liouville (RL) fractional derivatives and found the eigenfunctions as trigonometric and first-kind Bessel functions by using the method of separation of variables. Hasan et al. formulated an FOCP of distributed systems in spherical and cylindrical coordinates and obtained the numerical solutions by using Volterra-type integral equations [19].

Bisvas and Sen introduced a new numerical technique that can be used for problems in terms of RL or Caputo fractional derivatives [20]. In this method, a reflection operator is used to convert the right-sided fractional derivative into an equivalent left-sided term and then solves the FOCPs by a standard numerical method for solving FDEs. Another different numerical technique based on truncated fractional power series was produced in [21]. In this series, fractional powers were selected by taking into account the order of the fractional derivatives. In a similar thought, ordinary polynomials were used in [8].

The state-space representation of FOCPs in multi-dimensional spaces has also been studied in the literature. In [22], a multi-dimensional FOCP in which the state and control functions are in different dimensions. In a similar consideration, the same problem by changing the boundary conditions of state function and solving a different numerical example was presented [23]. A pseudo-state-space formulation for an FOCP in which a dynamic system involving both integer and fractional order derivatives was proposed [24].

Under different point of views, there are further studies on FOCPs as follows. Baleanu first studied the fractional variational problems in the presence of delay and so obtained the Euler–Lagrange equations [25]. Thus, the formulation of FOCPs with time delay was proposed in [26]. Tricaud and Chen in [27] and [28] produced a rational approximation for a large class of FOCPs, and so modelled the fractional dynamics of system in terms of a state-space form. Matignon formulated an optimal control problem of fractional linear systems by using adjoint systems to reduce the main problem into the state-space form [29].

Frederico and Torres introduced a Noether-type theorem for fractional optimal control systems in terms of RL [30] and Caputo [31] derivatives by using the Lagrange multiplier technique. In addition, they [32] proved a Noether-like theorem to the more general class of FOCPs. Jelicic and Petrovack formulated the necessary optimality conditions for optimal control problems whose dynamics are described by FDE with the highest integer-order derivative terms [33]. Rapaic and Jelicic investigated an optimal control problem, for a class of distributed parameter systems represent fractional heat conduction equation [34]. The existence of mild solutions for semi-linear fractional evolution equations, the Lagrange problem of systems governed by these equations and an existence result for optimal control problems were studied in [35]. The existence of time optimal control problem governed by fractional semi-linear evolution equations in terms of Caputo derivative was studied in [36]. These authors also focused on the numerical approximation of time optimal control problems whose dynamics are described by fractional evolution equations in Banach spaces [37]. Following the same manner, the existence of optimal control problem in which system dynamics are defined by fractional delay non-linear integro-differential equations in Banach spaces was studied [38]. Mophou consisted of two parts: the existence and uniqueness of solution of the fractional diffusion equation in Hilbert space constitutes the uniqueness of the optimal control problem governed by these equations [39].

So far, in the literature, the system dynamics of FOCPs have been considered as finite or infinite dimensional spaces; time-varying or unvarying types; linear or non-linear forms; with constraint or free terminal conditions, etc. However, for all these systems, only the effect of the time fractional derivatives in the sense of RL and Caputo have been analyzed, i.e. only the systems with time memory have been considered. However, it is well known that anomalous diffusion behaviours in the complex systems are defined by both time and space fractional operators. There is a question in our mind that 'How an FOCP of anomalous diffusion equation governed by time and space fractional derivatives can be formulated or solved?' This problem was first studied in one-dimensional Cartesian coordinates in [40].

In this chapter, we research the main problem [40] in spherical coordinates. The system is defined in terms of time Caputo and space fractional Laplacian operators. To obtain eigenfunctions, we use a spectral definition given for fractional Laplacian operators in bounded domain. The main advantage and the reason for choosing of this method is that spectral methods yield a fully diagonal matrix representation of the fractional operators and so have increased accuracy and efficiency when compared with low-order counterparts. In addition, these methods have a completely straightforward extension to two or three spatial dimensions in curvilinear coordinate systems as in the present study. The time discretization is completed by the Grünwald–Letnikov approach. The presented procedure is illustrated by a numerical example. It is shown by the figures that the limited number of state variables for computation is enough to convergence of the analytical and numerical solutions. In addition, the number of time grids for the time discretization is really desired small values. This significantly affects the working time of MATLAB® algorithm for such complicated FDE system. In this sense, it can be said that our numerical results are quite good. We also validate and show the effects of variable fractional orders on state and control functions.

The organization of this chapter is as follows. In Section 5.2, we give some fundamental definitions and mathematical relations necessary for our formulation. Section 5.3 has two subsections that present the spectral formulation of problem for half and complete axis symmetries, respectively. Section 5.4 explains the numerical results by the figures. Finally, we conclude the work in Section 5.5.

5.2 Preliminaries

Here, we briefly give some basic definitions and mathematical relations required for our problem formulation and obtaining its solution. We also mention the physical properties of the dynamic system on which an FOCP is considered.

In this chapter, we aim to formulate the FOCP for an anomalous diffusion process that was first introduced by Zaslavsky to model Hamiltonian chaos and also known as 'Time space-symmetric fractional differential equation (TSS-FDE)' in the literature [41]. The definition of a TSS-FDE in one-dimensional space is as follows:

$$ {}_tD^{\alpha}u(x,t) = -K_{\beta}\left(-\Delta\right)^{\frac{\beta}{2}} u(x,t), \quad 0 \le t \le T, \quad 0 \le x \le L, \tag{5.1} $$

where $0 < \alpha \le 1$ and $1 < \beta \le 2$ are the order of fractional Caputo derivative and fractional Laplacian operator. $K_{\beta} > 0$ represents the diffusion coefficient. Here, the symmetric space fractional derivative $-\left(-\Delta\right)^{\frac{\beta}{2}}$ of order β $(1 < \beta \le 2)$ is a fractional Laplacian operator defined through the eigenfunction expansion on a finite domain. Definition and the numerical computation techniques of this operator under the consideration of spectral representation were introduced in [42–45].

Definition 5.1 [46] Suppose the Laplacian $(-\Delta)$ has a complete set of orthonormal eigenfunctions φ_n corresponding to eigenvalues λ_n^2 on a bounded region D, i.e. $(-\Delta)\varphi_n = \lambda_n^2\varphi_n$ on D; $B(0) = 0$ on ∂D, where $B(\varphi)$ is one of the standard Dirichlet, Neumann and Robin boundary conditions. Let

$$ F_{\gamma} = \left\{ f = \sum_{n=1}^{\infty} c_n\varphi_n,\ c_n = \langle f, \varphi_n\rangle \middle| \sum_{n=1}^{\infty} |c_n|^2 |\lambda|_n^{\gamma} < \infty,\ \gamma = \max(\alpha, 0) \right\}. \tag{5.2} $$

Then for any $f \in F_{\gamma}$, $(-\Delta)^{\frac{\alpha}{2}}$ is defined by

$$ (-\Delta)^{\frac{\alpha}{2}} f = \sum_{n=1}^{\infty} c_n \left(\lambda_n\right)^{\alpha} \varphi_n. \tag{5.3} $$

In this work, we formulate the FOCP under a homogeneous Dirichlet boundary conditions. For a diffusion process, the physical behaviour of homogeneous Dirichlet boundary conditions means that the boundary is set far enough away from an evolving plume such that no significant concentrations reach that boundary. Another result under this assumption is that the fractional Laplacian operator equals to Riesz fractional derivative for one-dimensional infinite space [44], i.e. the spectral definition of a fractional Laplacian operator studied in the literature, using the Fourier transform on an infinite domain, is a natural extension to finite domain under the homogeneous Dirichlet boundary conditions.

The Caputo derivative is used to describe the time fractional effect in this work. It is well known that the Caputo fractional derivatives have been very popular as these operators allow someone to obtain physically interpretable initial conditions, i.e. the initial conditions for an FDE with Caputo time derivative are given in terms of integer-order derivatives.

If $f(t)$ is a time-dependent function and α $(n-1<\alpha<n)$ is the order of the derivative, the Caputo fractional derivatives are given as follows.

Definition 5.2 The left Caputo fractional derivative

$$ {}_{a}^{C}D_{t}^{\alpha} f(t) = \frac{1}{\Gamma(n-\alpha)} \int_{a}^{t} (t-\tau)^{n-\alpha-1} \left(\frac{d}{d\tau}\right)^{n} f(\tau)\, d\tau \tag{5.4} $$

and the right Caputo fractional derivative

$$ {}_{t}^{C}D_{b}^{\alpha} f(t) = \frac{1}{\Gamma(n-\alpha)} \int_{t}^{b} (\tau-t)^{n-\alpha-1} \left(-\frac{d}{d\tau}\right)^{n} f(\tau)\, d\tau. \tag{5.5} $$

To apply the GL approximation, we first consider the relation between RL and Caputo operators:

$$ {}_{a}^{RL}D_{x}^{\alpha} f(x) = {}_{a}^{C}D_{x}^{\alpha} f(x) + \sum_{k=0}^{n-1} \frac{d^k}{dx^k} f(x) \Big|_{x=a} \frac{(x-a)^{k-\alpha}}{\Gamma(k-\alpha+1)}, \tag{5.6} $$

$$ {}_{x}^{RL}D_{b}^{\alpha} f(x) = {}_{x}^{C}D_{b}^{\alpha} f(x) + \sum_{k=0}^{n-1} \frac{d^k}{dx^k} f(x) \Big|_{x=b} \frac{(b-x)^{k-\alpha}}{\Gamma(k-\alpha+1)} \tag{5.7} $$

where $n-1<\alpha<n$. The left and right RL fractional derivatives can be approximated by applying GL definition as

$$ {}_{0}^{RL}D_{t}^{\alpha} f \approx \frac{1}{h^{\alpha}} \sum_{j=0}^{M} w_j^{(\alpha)} f(hM-jh), \tag{5.8} $$

$$ {}_{t}^{RL}D_{1}^{\alpha} f \approx \frac{1}{h^{\alpha}} \sum_{j=0}^{N-M} w_j^{(\alpha)} f(hM+jh), \tag{5.9} $$

where

$$w_0^{(\alpha)} = 1, \quad w_j^{(\alpha)} = \left(1 - \tfrac{\alpha+1}{j}\right) w_{j-1}^{(\alpha)},$$

N is the number of subdomains that have $h = \frac{1}{N}$ lengths and M represents the nodes of time interval.

Problem: Find the optimal control $u(t)$ that minimizes the performance index

$$J(u) = \int_0^1 F(x, u, t)\, dt, \tag{5.10}$$

subject to dynamic constraints

$${}_0^C D_t^\alpha x = G(x, u, t), \tag{5.11}$$

and the initial conditions

$$x(0) = x_0, \tag{5.12}$$

where $x(t)$ and $u(t)$ are the state and control functions; F and G are arbitrary functions. By using calculus of variations and Lagrange multiplier technique, Euler–Lagrange equations which represent the necessary optimality conditions are defined as

$${}_0^C D_t^\alpha x = G(x, u, t), \tag{5.13}$$

$${}_t^C D_1^\alpha \lambda = \frac{\partial F}{\partial x} + \lambda \frac{\partial G}{\partial x}, \tag{5.14}$$

$$\frac{\partial F}{\partial u} + \lambda \frac{\partial G}{\partial u} = 0, \tag{5.15}$$

where λ is the Lagrange multiplier which is also known as costate variable, and initial state and final control values are

$$x(0) = x_0, \ \lambda(1) = 0. \tag{5.16}$$

The derivation of fractional Euler–Lagrange equations can be found in [8]. One can easily conclude from these equations that formulation of FOCPs requires both forward and backward fractional derivatives. This is the fundamental property that differentiates the FOCPs from the classic optimal control ones.

5.3 Formulation of Axis-Symmetric FOCP

In this section, we formulate the main FOCP for half and complete axis-symmetric cases in spherical coordinates. The objective of this work is to obtain an optimal control function that minimizes the quadratic performance index. For the half axis-symmetric case, the formulation is given as follows.

5.3.1 Half Axis-Symmetric Case

We aim to find the optimal control $u(t)$ for an anomalous diffusion system that minimizes the quadratic performance index

$$J(u) = \frac{1}{2}\int_0^1\int_0^\pi\int_0^a \left[Ax^2(r,\theta,t) + Bu^2(r,\theta,t)\right] r^2 \sin\theta drd\theta dt, \tag{5.17}$$

subject to the following system dynamics

$${}_t^C D^\alpha x(r,\theta,t) = -K_\beta(-\Delta)^{\frac{\beta}{2}} x(r,\theta,t) + u(r,\theta,t), \tag{5.18}$$

with the initial condition

$$x(r,\theta,0) = x_0(r,\theta), \tag{5.19}$$

and the homogeneous Dirichlet boundary conditions

$$x(a,\theta,t) = x(0,\theta,t) = 0, \tag{5.20}$$

where $0 < \alpha \leq 1$, $0 < \beta \leq 2$, $x(r,\theta,t)$ and $u(r,\theta,t)$ are the state and control functions, respectively; A and B are arbitrary constants; K_β denotes the anomalous diffusion coefficient which we assume $K_\beta > 0$, which means the flow is from left to right; ${}_t^C D^\alpha$ is the αth order of Caputo time fractional derivative; $-(-\Delta)^{\frac{\beta}{2}}$ notation represents the symmetric-space fractional derivative (i.e. fractional Laplacian operator).

Before obtaining the necessary optimality conditions, we first use the spectral definition of fractional Laplacian operator and so obtain eigenfunctions. Therefore, the state and control functions can be defined by a series of eigenfunctions. To obtain the eigenfunctions, let us consider the following Helmholtz equation

$$\Delta x(r,\theta) = -kx(r,\theta), \quad \begin{array}{l} 0 < r < a, \\ 0 < \theta < \pi, \end{array} \tag{5.21}$$

with the homogeneous boundary condition

$$x(a,\theta) = 0, \tag{5.22}$$

where the half-axis symmetric Laplace operator should be defined as

$$\Delta x(r,\theta) = \nabla^2 x(r,\theta) = \frac{\partial^2 x}{\partial r^2} + \frac{2}{r}\frac{\partial x}{\partial r} + \frac{1}{r^2}\left(\frac{\partial^2 x}{\partial \theta^2} + \cot\theta\frac{\partial x}{\partial \theta}\right). \tag{5.23}$$

It is a well-known task in the classical calculus to find the eigenvalues and eigenfunctions of this problem by the assumption $x(r,\theta) = R(r)Y(\theta)$. By applying separation of variables,

$$R_{n,m}(r) = j_n(\psi_{n,m} r), \quad n = 0,1,... \quad m = 1,2,... \tag{5.24}$$

where $\psi_{n,m} = \frac{\mu_{n+\frac{1}{2},m}}{a}$ and $\mu_{n+\frac{1}{2},m}$ is the mth positive zero of the Bessel function $J_{n+\frac{1}{2}}$ so that

$$j_n(r) = \left(\frac{\pi}{2r}\right)^{\frac{1}{2}} J_{n+\frac{1}{2}}(r). \tag{5.25}$$

Moreover, $j_n(r)$ denotes the nth-order spherical Bessel function. On the other hand, $Y(\theta)$ is obtained as

$$Y(\theta) = P_n(\cos\theta), \tag{5.26}$$

where P_n is Legendre polynomial of degree n. After all, the eigenvalues are $\psi_{n,m}$ such that $k = (\psi_{n,m})^2$ and the corresponding eigenfunctions are

$$x_{n,m}(r,\theta) = j_n(\psi_{n,m} r) P_n(\cos\theta), \quad n = 0, 1, ..., \quad m = 1, 2, ... \tag{5.27}$$

Let us remind that both j_n spherical Bessel functions and P_n Legendre polynomials satisfy the orthogonality property. Therefore, we should easily normalize these functions, and so we can represent the fractional Laplacian term in our system as a spectral form. The normalized form of eigenfunctions is as follows

$$x_{n,m}(r,\theta) = \left(\frac{2n+1}{2}\right)^{\frac{1}{2}} \frac{\sqrt{2}}{a\sqrt{a} j_{n+1}\left(\mu_{n+\frac{1}{2},m}\right)} j_n(\psi_{n,m} r) P_n(\cos\theta), \tag{5.28}$$

and $x_{n,m}(r,\theta)$ satisfies the homogeneous boundary condition

$$x_{n,m}(a,\theta) = 0, \quad 0 < \theta < \pi. \tag{5.29}$$

Therefore, we assume that the state and control functions take the following form

$$x(r,\theta,t) = \sum_{n=0}^{\infty} \sum_{m=1}^{\infty} x_{nm}(t) \left(\frac{2n+1}{2}\right)^{\frac{1}{2}} \frac{\sqrt{2}}{a\sqrt{a} j_{n+1}\left(\mu_{n+\frac{1}{2},m}\right)} \times j_n(\psi_{n,m} r) P_n(\cos\theta), \tag{5.30}$$

and

$$u(r,\theta,t) = \sum_{n=0}^{\infty} \sum_{m=1}^{\infty} u_{nm}(t) \left(\frac{2n+1}{2}\right)^{\frac{1}{2}} \frac{\sqrt{2}}{a\sqrt{a} j_{n+1}\left(\mu_{n+\frac{1}{2},m}\right)} \times j_n(\psi_{n,m} r) P_n(\cos\theta). \tag{5.31}$$

In addition, using Definition 5.1, the fractional Laplacian operator for $x(r,\theta,t)$ is given as

$$(-\Delta)^{\frac{\beta}{2}} x(r,\theta,t) = \sum_{n=0}^{\infty} \sum_{m=1}^{\infty} x_{nm}(t) (\psi_{n,m})^{\beta} x_{n,m}(r,\theta). \tag{5.32}$$

By substituting Eqs. (5.30)–(5.32) into Eqs. (5.17) and (5.18), F and G functions mentioned in the general definition of an FOCP can be calculated as

$$F_{nm}(x,u,t) = \frac{1}{2}\left[Ax_{nm}^2(t) + Bu_{nm}^2(t)\right], \tag{5.33}$$

$$G_{nm}(x,u,t) = -(\psi_{n,m})^{\beta} x_{nm}(t) + u_{nm}(t), \tag{5.34}$$

and so we can obtain the necessary optimality conditions by using Eqs. (5.13)–(5.15) as follows

$${}_{0}^{C}D_{t}^{\alpha} x_{nm}(t) = -(\psi_{n,m})^{\beta} x_{nm}(t) + u_{nm}(t), \tag{5.35}$$

$${}_{t}^{C}D_{1}^{\alpha} \lambda_{nm}(t) = Ax_{nm}(t) - (\psi_{n,m})^{\beta} \lambda_{nm}(t), \tag{5.36}$$

$$Bu_{nm}(t) + \lambda_{nm}(t) = 0, \tag{5.37}$$

where $\lambda_{nm}(t)$ denotes the Lagrange multiplier (i.e. costate variable). After rearrangement of these equations, we get

$${}_{0}^{C}D_{t}^{\alpha} x_{nm}(t) = -(\psi_{n,m})^{\beta} K_{\beta} x_{nm}(t) + u_{nm}(t), \tag{5.38}$$

$${}_{t}^{C}D_{1}^{\alpha} u_{nm}(t) = -\frac{A}{B} x_{nm}(t) - (\psi_{n,m})^{\beta} K_{\beta} u_{nm}(t), \tag{5.39}$$

with the terminal conditions

$$x_{nm}(0) = x_{nm}(r,\theta), \tag{5.40}$$

$$u_{nm}(1) = 0. \tag{5.41}$$

To solve the FDE system (5.38–5.39), we need to know the initial condition function $x_{nm}(0)$. For this purpose, we use the following steps:

1. Assume $t = 0$ in Eq. (5.30),
2. Multiply both sides of the equation by $P_n(\cos\theta)\sin\theta$ and then integrate 0 to π,
3. Multiply both sides of the equation by $j_n(\psi_{n,m} r) r^2$ and then integrate 0 to a,
4. Use the orthogonality properties of functions P_n and j_n, and obtain the initial condition function $x_{nm}(0)$ as follows:

$$x_{nm}(0) = \frac{(2n+1)^{1/2}}{a\sqrt{a}\, j_{n+1}\left(\mu_{n+\frac{1}{2},m}\right)} \int_0^{\pi}\int_0^{a} x(r,\theta,0)\, P_n(\cos\theta)\sin\theta\, j_n(\psi_{n,m} r)\, r^2 d\theta dr. \tag{5.42}$$

Note that $x_{nm}(0)$ depends on the choice of $x(r,\theta,0)$, that is initial state condition of the problem. We arbitrarily choose

$$x(r,\theta,0) = P_0(\cos\theta)\, j_0(\psi_{0,1} r), \tag{5.43}$$

and so obtain

$$x_{nm}(0) = \begin{cases} j_1\left(\mu_{\frac{1}{2},1}\right); & n = 0,\ m = 1 \\ 0 & n > 0, m > 1. \end{cases} \quad (5.44)$$

We also remind that final value of control function is $u_{nm}(1) = 0$. To obtain the analytical solutions, we get $\alpha = 1$ in Eqs. (5.38) and (5.39), and the solutions are as follows

$$x_{nm}(t) = x_{nm}(0)\left[\frac{b_{nm}\cosh(b_{nm}(t-1)) - a_{nm}\sinh(b_{nm}(t-1))}{b_{nm}\cosh(b_{nm}) + a_{nm}\sinh(b_{nm})}\right], \quad (5.45)$$

$$u_{nm}(t) = x_{nm}(0)\left[\frac{\left(b_{nm}^2 - a_{nm}^2\right)\sinh(b_{nm}(t-1))}{b_{nm}\cosh(b_{nm}) + a_{nm}\sinh(b_{nm})}\right]. \quad (5.46)$$

where

$$a_{nm} = K_\beta(\psi_{n,m})^\beta, \quad (n = 0, 1, ...\Lambda\ m = 1, 2, ...),$$

$$b_{nm} = \sqrt{\frac{A}{B} + a_{nm}^2}.$$

The systems (5.38) and (5.39) are approximated by using the GL approximation given by Eqs. (5.6) and (5.7) as

$$\begin{aligned} &\frac{1}{h^\alpha}\sum_{j=0}^{M} w_j^{(\alpha)} x_{nm}(Mh - jh) - x_{nm}(0)\frac{[Mh]^{-\alpha}}{\Gamma(1-\alpha)} \\ &+ K_\beta(\psi_{n,m})^\beta x_{nm}(Mh) = u_{nm}(Mh) \\ &(M = 1, 2, ..., N, \quad n = 0, 1, ..., p, \quad m = 1, 2, ..., q), \end{aligned} \quad (5.47)$$

$$\begin{aligned} &\frac{1}{h^\alpha}\sum_{j=0}^{N-M} w_j^{(\alpha)} x_{nm}(Mh + jh) - u_{nm}(Nh)\frac{[(N-M)h]^{-\alpha}}{\Gamma(1-\alpha)} \\ &+ K_\beta(\psi_{n,m})^\beta u_{nm}(Mh) = -\frac{A}{B}x_{nm}(Mh) \\ &(M = N-1, N-2, ..., 0, \quad n = 0, 1, ..., p, \quad m = 1, 2, ..., q). \end{aligned} \quad (5.48)$$

Now, we will similarly formulate the complete axis-symmetric case.

5.3.2 Complete Axis-Symmetric Case

Let us consider the following performance index

$$J(u) = \frac{1}{2}\int_0^1\int_0^a \left[Ax^2(r,t) + Bu^2(r,t)\right] r^2 dr dt, \quad (5.49)$$

subject to the dynamic system

$$ {}_{t}^{C}D^{\alpha}x(r,t) = -K_{\beta}(-\Delta)^{\frac{\beta}{2}}x(r,t) + u(r,t), \tag{5.50} $$

where the initial and homogeneous boundary conditions are given as

$$ x(r,0) = x_0(r), \tag{5.51} $$

$$ x(a,t) = x(0,t) = 0. \tag{5.52} $$

By similar computations to Part 1, we obtain the eigenvalues

$$ \psi_m = \frac{\mu_{\frac{1}{2},m}}{a}, \quad m = 1, 2, ... \tag{5.53} $$

where $\mu_{\frac{1}{2},m}$ $(m = 1, 2, ...)$ are the positive zeros of $J_{\frac{1}{2}}$ Bessel function, and the related eigenfunctions are

$$ R_m(r) = j_0(\psi_m r). \tag{5.54} $$

Orthonormalized forms of eigenfunctions are found as

$$ R_m(r) = \frac{\sqrt{2}}{a\sqrt{a}j_1\left(\mu_{\frac{1}{2},m}\right)}j_0(\psi_m r). $$

Now, it is possible to assume that the state and control functions can be represented by a series that is composed of the linear form of orthonormalized form of eigenfunctions as follows

$$ x(r,t) = \sum_{m=1}^{\infty} x_m(t)\frac{\sqrt{2}}{a\sqrt{a}j_1\left(\mu_{\frac{1}{2},m}\right)}j_0(\psi_m r), \tag{5.55} $$

$$ u(r,t) = \sum_{m=1}^{\infty} u_m(t)\frac{\sqrt{2}}{a\sqrt{a}j_1\left(\mu_{\frac{1}{2},m}\right)}j_0(\psi_m r). \tag{5.56} $$

We recalculate the necessary optimality equations and reorganize the Equations, and then we take the main FDEs for the state and control eigencoordinate functions as

$$ {}_{0}^{C}D_{t}^{\alpha}x_m(t) = -K_{\beta}(\psi_m)^{\beta}x_m(t) + u_m(t), \tag{5.57} $$

$$ {}_{t}^{C}D_{1}^{\alpha}u_m(t) = -\frac{A}{B}x_m(t) - (\psi_m)^{\beta}K_{\beta}u_m(t), \; m = 1, 2, ... \tag{5.58} $$

In a similar manner to the earlier steps to find the initial condition function, the following result is obtained

$$ x_m(0) = \frac{\sqrt{2}}{a\sqrt{a}j_1\left(\mu_{\frac{1}{2},m}\right)}\int_0^a x_0(r)j_0(\psi_m r)r^2dr. $$

For $\alpha = 1$, the analytical solutions of the system (5.57)–(5.58) are as follows

$$x_m(t) = c_1 \exp(-p(m)t) + c_2 \exp(p(m)t), \tag{5.59}$$

$$u_m(t) = c_1 (q(m) - p(m)) \exp(-p(m)t) + c_2 (q(m) + p(m)) \exp(p(m)t), \tag{5.60}$$

where

$$q(m) = K_\beta (\psi_m)^\beta,$$
$$p(m) = \sqrt{q^2(m) + 1},$$
$$c_1 = \frac{(p(m) - q(m))}{\{(p(m) - q(m)) + (p(m) + q(m)) \exp(2p(m))\}} x_m(0),$$
$$c_2 = \frac{(p(m) + q(m)) \exp(2p(m))}{\{(p(m) - q(m)) + (p(m) + q(m)) \exp(2p(m))\}} x_m(0).$$

To write the numerical algorithms in MATLAB and obtain numerical solutions, we use the following approximated form of system (5.57)–(5.58)

$$\begin{aligned}
&\frac{1}{h^\alpha} \sum_{j=0}^{M} w_j^{(\alpha)} x_m(Mh - jh) - x_m(0) \frac{[Mh]^{-\alpha}}{\Gamma(1-\alpha)} \\
&+ K_\beta (\psi_m)^\beta x_m(Mh) = u_m(Mh) \\
&(M = 1, 2, ..., N, \quad m = 0, 1, ..., p),
\end{aligned}$$

$$\begin{aligned}
&\frac{1}{h^\alpha} \sum_{j=0}^{N-M} w_j^{(\alpha)} x_m(Mh + jh) - u_m(Nh) \frac{[(N-M)h]^{-\alpha}}{\Gamma(1-\alpha)} \\
&+ K_\beta (\psi_m)^\beta u_m(Mh) = -\frac{A}{B} x_m(Mh) \\
&(M = N-1, N-2, ..., 0 \quad m = 0, 1, ..., p).
\end{aligned}$$

In the next section, we will analyze our formulation in the sense of analytical and numerical for the half-axis symmetric case, and we also validate the effects of variations of problem parameters under the choice of initial condition (5.44). We remind that complete-axis symmetry is a special case of the first part so that if we take $\theta = 0$ in the algorithms of numerical calculations, we can see that similar results occur.

5.4 Numerical Results

In this section, we give the numerical results by MATLAB figures. Here, figures correspond to $x_{01}(t)$ and $u_{01}(t)$ functions because of our arbitrary choice of initial condition. Here we examine the $x_{01}(t)$ state and $u_{01}(t)$ control component functions that contribute mostly to the series solutions (5.30) and (5.31).

First, we compare the analytical and GL numerical solutions for $\beta = 1.75$ and $N = 100$ in Figure 5.1a and b and also show the effect of variation of α parameter on the component functions. As is evident from these figures, while $\alpha \to 1$ and $\beta \to 2$, the system behaviour approaches to the classical diffusion.

In Figure 5.2a and b we fix α time-order parameter and analyze the change of space order β. In the figure, we see that the effect of variation of the space fractional order β is less than the effect of variation of α. We see that time

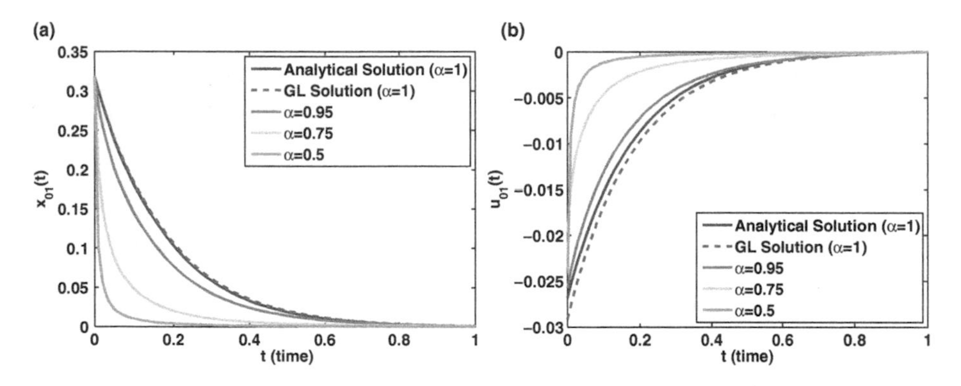

FIGURE 5.1
Behaviours of (a) $x_{01}(t)$ state and (b) $u_{01}(t)$ control eigencoordinate functions for different values of α: $\beta = 1.75$ and $N = 100$.

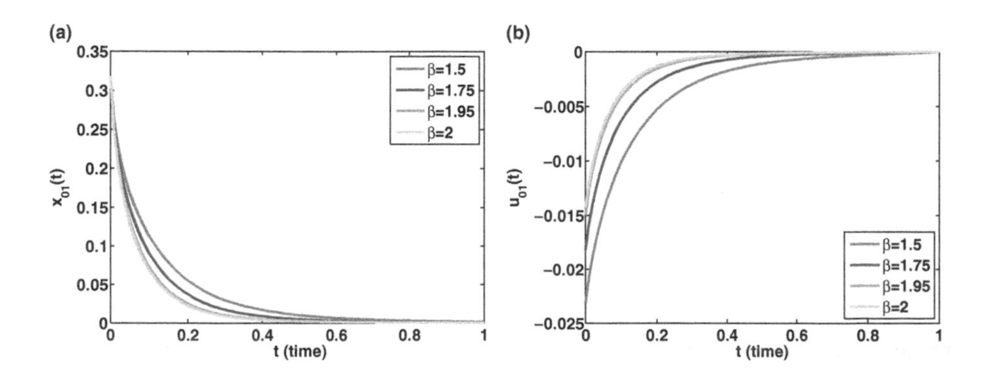

FIGURE 5.2
Behaviours of (a) $x_{01}(t)$ state and (b) $u_{01}(t)$ control eigencoordinate functions for different values of β: $\alpha = 0.9$ and $N = 100$.

step number 50 for discretization has been enough to convergence of solutions. (Figure 5.3a and b). This number is quite enough for the running time of MATLAB algorithms. This result also reveals the advantage of the preference of GL approach.

We plot the whole solutions $x(r,t)$ and $u(r,t)$ of the problem in Figure 5.4a and b for the fixed value $\theta = \frac{\pi}{3}$ and we also take $\alpha = 0.9, \beta = 1.75$ and $N = 100$. Similarly, we show the solutions $x(\theta,t)$ and $u(\theta,t)$ for the values $r = 0.5, \alpha = 0.9, \beta = 1.75$ and $N = 100$ in Figure 5.5a and b. Note that all of these choices of the parameters for surfaces are arbitrary.

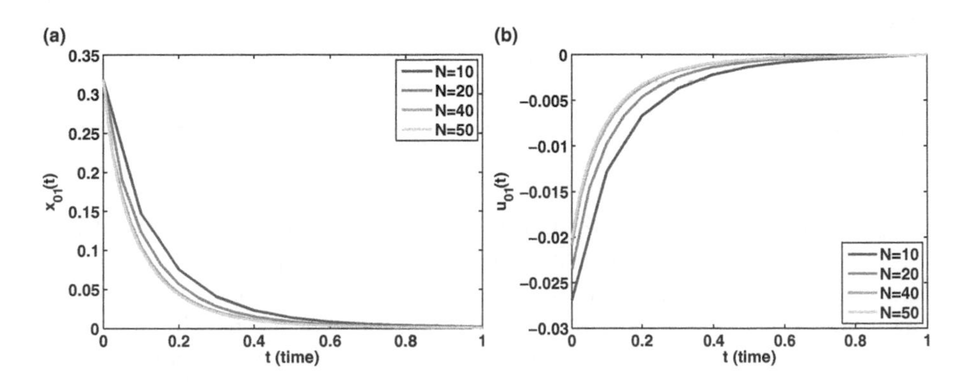

FIGURE 5.3
Effect of step number N on the convergence of eigencoordinate functions (a) $x_{01}(t)$ state and (b) $u_{01}(t)$ for $\alpha = 0.9$, $\beta = 1.75$.

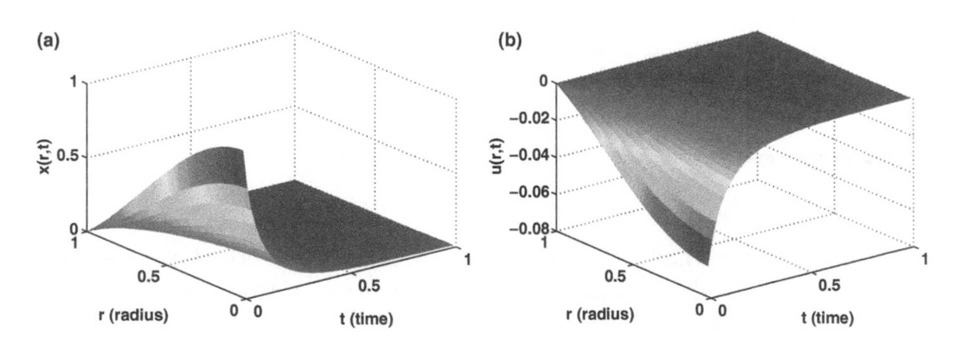

FIGURE 5.4
Surface plots of (a) $x(r,t)$ and (b) $u(r,t)$ for $\alpha = 0.9$, $\beta = 1.75$, $N = 100$, $\theta = \Pi/3$ and $n = m = 4$.

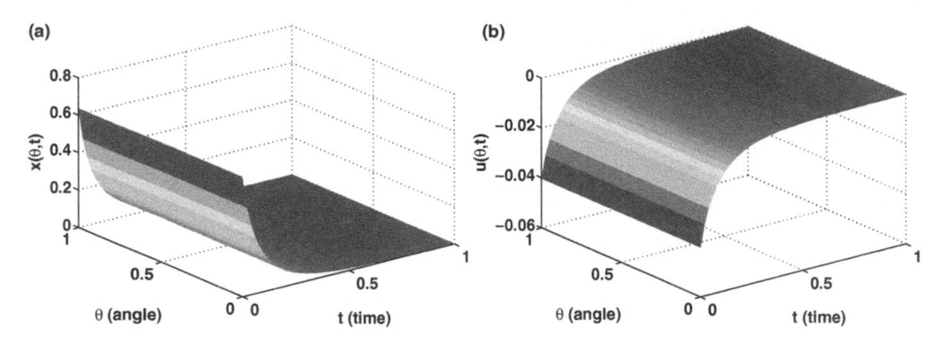

FIGURE 5.5
Surface plots of (a) $x(\theta, t)$ and (b) $u(\theta, t)$ for $\alpha = 0.9$, $\beta = 1.75$, $N = 100$, $r = 0.5$ and $n = m = 4$.

5.5 Conclusions

In this chapter, an optimal control problem of a system defined by a TSS-FDE is considered in spherical coordinates and for half and complete axis-symmetric cases. The time and space fractional derivatives are in terms of Caputo and fractional Laplacian operators. Spectral representation for fractional Laplacian operator is used to discrete the space term of the system and control functions. Lagrange multiplier technique is applied to calculate the necessary optimality conditions. To discrete the time, GL approximation is applied. The effectiveness of the numerical method is ensured by the comparison of analytical and numerical solutions. Figures are used to show the behaviour of the problem under the variation of problem parameters. By this study, not only the fractional time effect to the dynamic system defined by a fractional diffusion-wave equation but also the effect of space fractional term on the system and control is considered. Therefore, this work generalizes the dynamic system structure by considering both fractional time and space effects.

References

[1] Akgül, A. and Khan, Y. (2017). A novel simulation methodology of fractional order nuclear science model. *Mathematical Methods in the Applied Sciences*, 40(17), 6208–6219.

[2] Singh, H., Srivastava, H. M. and Kumar, D. (2017). A reliable numerical algorithm for the fractional vibration equation. *Chaos, Solitons and Fractals*, 103, 131–138.

[3] Yavuz, M. and Özdemir, N. (2019). Comparing the new fractional derivative operators involving exponential and Mittag-Leffler kernel. *Discrete and Continuous Dynamical Systems-S*, 1098–1107.

[4] Singh, C. S., Singh, H. Singh, V. K. and Singh, O. P. (2016). Fractional order operational matrix methods for fractional singular integro-differential equation. *Applied Mathematical Modelling*, 40(23–24), 10705–10718.

[5] Akgül, A., Karatas, E. and Baleanu, D. (2015). Numerical solutions of fractional differential equations of Lane-Emden type by an accurate technique. *Advances in Difference Equations*, 2015(1), 220.

[6] Yavuz, M. (2019). Characterizations of two different fractional operators without singular kernel. *Mathematical Modelling of Natural Phenomena*, 14(3), 302.

[7] Singh, H., Srivastava, H. M. and Kumar, D. (2018). A reliable algorithm for the approximate solution of the nonlinear Lane-Emden type equations arising in astrophysics. *Numerical Methods for Partial Differential Equations*, 34(5), 1524–1555.

[8] Agrawal, O. P. (2004). A general formulation and solution scheme for fractional optimal control problems. *Nonlinear Dynamics*, 38(1–4), 323–337.

[9] Agrawal, O. P. (2008). A formulation and numerical scheme for fractional optimal control problems. *Journal of Vibration and Control*, 14(9–10), 1291–1299.

[10] Agrawal, O. P. (2008). A quadratic numerical scheme for fractional optimal control problems. *Journal of Dynamic Systems, Measurement, and Control*, 130(1), 011010.

[11] Tangpong, X. W. and Agrawal, O. P. (2009). Fractional optimal control of continuum systems. *Journal of Vibration and Acoustics*, 131(2), 021012.

[12] Agrawal, O. P. and Baleanu, D. (2007). A Hamiltonian formulation and a direct numerical scheme for fractional optimal control problems. *Journal of Vibration and Control*, 13(9–10), 1269–1281.

[13] Baleanu, D., Defterli, O. and Agrawal, O. P. (2009). A central difference numerical scheme for fractional optimal control problems. *Journal of Vibration and Control*, 15(4), 583–597.

[14] Biswas, R. K. and Sen, S. (2011). Fractional optimal control problems with specified final time. *Journal of Computational and Nonlinear Dynamics*, 6(2), 021009.

[15] Agrawal, O. P. (2008). Fractional optimal control of a distributed system using eigenfunctions. *Journal of Computational and Nonlinear Dynamics*, 3(2), 021204.

[16] Özdemir, N., Agrawal, O. P., İskender, B. B. and Karadeniz, D. (2009). Fractional optimal control of a 2-dimensional distributed system using eigenfunctions. *Nonlinear Dynamics*, 55(3), 251.

[17] Özdemir, N., Agrawal, O. P., Karadeniz, D. and Iskender, B. B. (2009). Fractional optimal control problem of an axis-symmetric diffusion-wave propagation. *Physica Scripta*, 2009(T136), 014024.

[18] Özdemir, N., Karadeniz, D. and Iskender, B. B. (2009). Fractional optimal control problem of a distributed system in cylindrical coordinates. *Physics Letters A*, 373(2), 221–226.

[19] Hasan, M. M., Tangpong, X. W. and Agrawal, O. P. (2012). Fractional optimal control of distributed systems in spherical and cylindrical coordinates. *Journal of Vibration and Control*, 18(10), 1506–1525.

[20] Biswas, R. K. and Sen, S. (2009). Numerical method for solving fractional optimal control problems. In *ASME 2009 International Design Engineering Technical Conferences and Computers and Information in Engineering Conference* (pp. 1205–1208). American Society of Mechanical Engineers. doi: 10.1115/DETC2009-87008.

[21] Agrawal, O. P. (2009). A numerical scheme and an error analysis for a class of Fractional Optimal Control problems. In *ASME 2009 International Design Engineering Technical Conferences and Computers and Information in Engineering Conference* (pp. 1253–1260). American Society of Mechanical Engineers.

[22] Agrawal, O. P., Defterli, O. and Baleanu, D. (2010). Fractional optimal control problems with several state and control variables. *Journal of Vibration and Control*, 16(13), 1967–1976.

[23] Defterli, O. (2010). Corrigendum to "A numerical scheme for two dimensional optimal control problems with memory effect" [*Comput. Math. Appl.* 59 (2010), 1630–1636]. *Computers and Mathematics with Applications*, 59(8), 3028.

[24] Biswas, R. K. and Sen, S. (2011). Fractional optimal control problems: a pseudo-state-space approach. *Journal of Vibration and Control*, 17(7), 1034–1041.

[25] Baleanu, D., Maaraba, T. and Jarad, F. (2008). Fractional variational principles with delay. *Journal of Physics A: Mathematical and Theoretical*, 41(31), 315403.

[26] Jarad, F., Abdeljawad, T. and Baleanu, D. (2010). Fractional variational optimal control problems with delayed arguments. *Nonlinear Dynamics*, 62(3), 609–614.

[27] Tricaud, C. and Chen, Y. (2009, June). Solution of fractional order optimal control problems using SVD-based rational approximations. In *2009 American Control Conference* (pp. 1430–1435). IEEE.

[28] Tricaud, C. and Chen, Y. (2010). An approximate method for numerically solving fractional order optimal control problems of general form. *Computers and Mathematics with Applications*, 59(5), 1644–1655.

[29] Matignon, D. (2010) Optimal control of fractional systems: A diffusive formulation. In *19th International Symposium on Mathematical Theory of Networks and Systems-MTNS 2010*, Budapest, Hungary, 05-09 Jul 2010.

[30] Frederico, G. S. and Torres, D. F. (2006). Noether's theorem for fractional optimal control problems. *IFAC Proceedings Volumes*, 39(11), 79–84.

[31] Frederico, G. S. and Torres, D. F. (2008). Fractional optimal control in the sense of caputo and the fractional Noether's theorem. *International Mathematical Forum*, 3(10), 479–493.

[32] Frederico, G. S. and Torres, D. F. (2008). Fractional conservation laws in optimal control theory. *Nonlinear Dynamics*, 53(3), 215–222.

[33] Jelicic, Z. D. and Petrovacki, N. (2009). Optimality conditions and a solution scheme for fractional optimal control problems. *Structural and Multidisciplinary Optimization*, 38(6), 571–581.

[34] Rapaić, M. R. and Jeliŏić, Z. D. (2010). Optimal control of a class of fractional heat diffusion systems. *Nonlinear Dynamics*, 62(1–2), 39–51.

[35] Wang, J. and Zhou, Y. (2011). A class of fractional evolution equations and optimal controls. *Nonlinear Analysis: Real World Applications*, 12(1), 262–272.

[36] Wang, J. and Zhou, Y. (2011). Time optimal control problem of a class of fractional distributed systems. *International Journal of Dynamical Systems and Differential Equations*, 3(3), 363–382.

[37] Wang, J. and Zhou, Y. (2011). Study of an approximation process of time optimal control for fractional evolution systems in Banach spaces. *Advances in Difference Equations*, 2011(1), 385324.

[38] Wang, J., Zhou, Y. and Wei, W. (2011). A class of fractional delay nonlinear integrodifferential controlled systems in Banach spaces. *Communications in Nonlinear Science and Numerical Simulation*, 16(10), 4049–4059.

[39] Mophou, G. M. (2011). Optimal control of fractional diffusion equation. *Computers and Mathematics with Applications*, 61(1), 68–78.

[40] Özdemir, N. and Avci, D. (2014). Optimal control of a linear time-invariant space–time fractional diffusion process. *Journal of Vibration and Control*, 20(3), 370–380.

[41] Zaslavsky, G. M. (1994). Fractional kinetic equation for Hamiltonian chaos. *Physica D: Nonlinear Phenomena*, 76(1–3), 110–122.

[42] Ilic, M., Liu, F., Turner, I. and Anh, V. (2005). Numerical approximation of a fractional-in-space diffusion equation, I. *Fractional Calculus and Applied Analysis*, 8(3), 323–341.

[43] Yang, Q., Turner, I. and Liu, F. (2009). Analytical and numerical solutions for the time and space-symmetric fractional diffusion equation. *ANZIAM Journal*, 50, 800–814.

[44] Yang, Q., Liu, F. and Turner, I. (2010). Numerical methods for fractional partial differential equations with Riesz space fractional derivatives. *Applied Mathematical Modelling*, 34(1), 200–218.

[45] Shen, S., Liu, F. and Anh, V. (2011). Numerical approximations and solution techniques for the space-time Riesz–Caputo fractional advection-diffusion equation. *Numerical Algorithms*, 56(3), 383–403.

[46] Ilic, M., Liu, F., Turner, I. and Anh, V. (2006). Numerical approximation of a fractional-in-space diffusion equation (II)–with nonhomogeneous boundary conditions. *Fractional Calculus and Applied Analysis*, 9(4), 333–349.

6

Integral-Balance Methods for the Fractional Diffusion Equation Described by the Caputo-Generalized Fractional Derivative

Ndolane Sene and Abdon Atangana

Université Cheikh Anta Diop de Dakar and University of the Free State

CONTENTS

6.1 Introduction

The analytical solutions, the numerical solutions and the approximate solutions are the subject of many investigations, in the problem consisting of getting the solutions of the fractional differential equations. In this chapter, we are interested in the approximate solutions of a particular class of diffusion equations. We use the integral-balance methods of getting the approximate solution of the fractional diffusion equations. The integral-balance method includes the heat balance integral method (HBIM) [7, 19] and the double integral method (DIM) [7,19].

Many investigations related to the HBIM and the DIM exist in the literature. In [7], Hristov has used the HBIM and the DIM for proposing an approximate solution of the fractional space diffusion. In [6], Hristov has suggested the HBIM and the DIM of getting an approximate solution of the non-linear Dodson diffusion equation. He illustrates the main results by computing the approximate solutions. In [10], the author has proposed an explicit solution of the heat radiation diffusion equation using the integral balance method. The HBIM and the DIM can be applied in the Stokes problems [12], in the fluid equations [12], in the diffusion equations [3, 18, 19], in the Stefan problem [17, 19], in the fourth-order fractional diffusion equation [8], in the Mullin models [5] and many others.

The HBIM and the DIM are known to be very practical in the problem consisting of getting an approximate solution of a particular class of diffusion equations. They use physical concepts. The integral-balance method includes assuming that the profile of the diffusion equation is polynomial. We will determine the finite penetration depth of the assumed profile. The integral approach is popular but has some limitations in its applicability. The determination of the exponent n of the assumed profile is not trivial. In [19], Myer and Mitchell propose an optimization method of getting this exponent n with the diffusion equation in integer order. This same approach is used for getting the exponent n in non-integer order [11]. In the literature, there exist many works related to the exponent n. In many problems, it was proved that it is better to choose the exponents $n = 2$ and $n = 3$ [19]. In general, the exponent candidate n is into the interval $[2, 3]$. In this chapter, we accept this conjecture.

Recently, the generalized fractional derivative and ρ-Laplace transform were introduced in the literature [13]. They open a new door in fractional calculus. There exist many fractional derivative operators in the research of fractional calculus. All the fractional derivative operators are equivalent. We have the Caputo fractional derivative [16,21,22], the Riemann–Liouville fractional derivative [16,21], the Caputo–Fabrizio fractional derivative [2,4,25], the Atangana–Baleanu fractional derivative [1,24], the Conformable derivative [20, 23], certain generalized fractional derivative operators and others [13,14]. Our chapter addresses the effect of the second-order ρ of the Caputo-generalized fractional derivative in the diffusion process. We mainly investigate the impact of ρ in the behaviour of the assumed profile. We consider the exponents of the assumed profile are $n = 2$ and $n = 3$, respectively [19]. We propose according to Myers and Mitchell optimization method the way of getting the exponent n for an accurate assumed profile. We are motivated by the new open problems generated by the Caputo-generalized fractional derivative. The question consists of getting the analytical solution of the fractional diffusion equation described by the Caputo-generalized fractional derivative.

The organization of this chapter is as follows. In Section 6.2, we recall the fractional calculus news. In Section 6.3, we give some basic calculations used for the integral-balance methods. In Section 6.4, we describe the integral-balance method for the generalized fractional diffusion equation represented by the Caputo fractional derivative. In Section 6.5, the approximate solution of the generalized fractional diffusion equation is investigated. In Section 6.6, we propose and describe Myers and Mitchell approach of getting the exponent n for the assumed profile. In Section 6.7, we give the concluding remarks.

6.2 Fractional Calculus News

In this section, we recall the necessary tools for this chapter. Fractional calculus is a field of mathematics in which we use the non-integer order derivative operators. The fractional derivative operators are the Riemann–Liouville derivative, the Caputo derivative, the Atangana–Baleanu derivative, the Caputo-Fabrizio derivative, and many others. In this section, we are interested in the generalized form of the non-integer derivative operators, namely the Riemann–Liouville generalized fractional derivative and the Caputo-generalized fractional derivative. We begin this section by recalling the generalized fractional integral which we use to define the generalized fractional derivative operators.

For a given function $f : [a, +\infty[\longrightarrow \mathbb{R}$, the generalized fractional integral [13,15] of order α, $\rho > 0$ of the function f is expressed in the following form:

$$\left(I^{\alpha,\rho} f\right)(t) = \frac{\rho^{1-\alpha}}{\Gamma(\alpha)} \int_a^t \left(t^\rho - s^\rho\right)^{\alpha-1} f(s) \frac{ds}{s^{1-\rho}}, \tag{6.1}$$

for all $t > a$, $0 < \alpha < 1$ and $\Gamma(.)$ is the Gamma function. We observe that when the order $\rho = 1$, we recover the classical fractional integral defined by

$$\left(I^{\alpha} f\right)(t) = \frac{1}{\Gamma(\alpha)} \int_a^t (t - s)^{\alpha-1} f(s) ds. \tag{6.2}$$

For a given function $f : [a, +\infty[\longrightarrow \mathbb{R}$, the generalized fractional derivative in Riemann–Liouville sense [13,15] of order α, $\rho > 0$ of the function f is defined as the following form:

$$\left({}_aD^{\alpha,\rho} f\right)(t) = \frac{\gamma^n}{\Gamma(n-\alpha)} \int_a^t \left(\frac{t^\rho - s^\rho}{\rho}\right)^{n-\alpha-1} f(s) \frac{ds}{s^{1-\rho}}, \tag{6.3}$$

for all $t > a$, where $n - 1 < \alpha < n$ and $\gamma = t^{1-\rho}\frac{d}{dt}$. We can also notice when the order $\rho = 1$, we recover the classical Riemann–Liouville fractional derivative expressed as

$$\left({}_aD^{\alpha} f\right)(t) = \frac{1}{\Gamma(n-\alpha)} \left(\frac{d}{dt}\right)^n \int_a^t (t - s)^{n-\alpha-1} f(s) ds. \tag{6.4}$$

For a given function $f : [a, +\infty[\longrightarrow \mathbb{R}$, the Caputo-generalized fractional derivative [13,15] of order α, $\rho > 0$ of f is represented in the following form:

$$\left({}_aD_c^{\alpha,\rho} f\right)(t) = \left({}_aI^{n-\alpha,\rho}\gamma^n f\right)(t) \qquad n = \lfloor \alpha \rfloor + 1. \tag{6.5}$$

We observe that when the order $\rho = 1$, we recover the classical Caputo fractional derivative defined as the following form:

$${}_aD_c^{\alpha} f(t) = \frac{1}{\Gamma(n-\alpha)} \int_a^t (t-s)^{n-\alpha-1} f^{(n)}(s)ds, \tag{6.6}$$

for all $t > a$ and $n - 1 < \alpha < n$.

We finish this section by recalling the ρ-Laplace transform of the Caputo-generalized fractional derivative recently introduced in the literature by Fahd and Thabet in [13]. We have the following identity:

$$\mathcal{L}_\rho \left\{ \left({}_aD^{\alpha,\rho} f\right)(t) \right\} = s^\alpha \mathcal{L}_\rho \left\{ f(t) \right\} - \sum_{k=0}^{n-1} s^{\alpha-k-1} \left(\gamma^n f\right)(0) \tag{6.7}$$

For the other fractional derivative operators which we have not defined in this section, see the literature. $\mathcal{L}_\rho$ designs the usual ρ-Laplace transform [13].

6.3 Basics Calculus for the Integral-Balance Methods

In this section, we recall some strategical calculations fundamental for the integral-balance methods: the HBIM and DIM. The assumed profile for the integral-balance methods is defined as the following form [10]:

$$u(x,t) = U_0 \left(1 - \frac{x}{\delta}\right)^n, \tag{6.8}$$

where δ represents the finite penetration depth. For the rest of this chapter, we consider the assumed profile defined as follows:

$$\bar{u}(x,t) = \frac{u_a(x,t)}{U_0} = \left(1 - \frac{x}{\delta}\right)^n. \tag{6.9}$$

Let's recall some fundamental properties for the HBIM and the DIM. For the HBIM, we integrate the fractional diffusion equation between 0 to the finite penetration depth δ. We have the following results:

$$\bullet\bullet\bullet \int_0^\delta D_t^{\alpha,\rho}\bar{u}(x,t)dx = D^{\alpha,\rho}\int_0^\delta \left(1-\frac{x}{\delta}\right)^n dx$$
$$= D_t^{\alpha,\rho}\left[-\frac{\delta}{n+1}\left(1-\frac{x}{\delta}\right)^{n+1}\right]_0^\delta$$
$$= D_t^{\alpha,\rho}\left[\frac{\delta}{n+1}\right]$$
$$= \frac{1}{n+1}D_t^{\alpha,\rho}\delta. \tag{6.10}$$

$$\bullet\bullet\bullet \int_0^\delta \frac{\partial^2\bar{u}}{\partial x^2}dx = \left[\frac{\partial\bar{u}}{\partial x}\right]_0^\delta$$
$$= -\frac{\partial\bar{u}}{\partial x}_{\|x=0}$$
$$= \frac{n}{\delta}\left(1-\frac{x}{\delta}\right)^{n-1}_{\|x=0}$$
$$= \frac{n}{\delta}. \tag{6.11}$$

For the DIM, we use double integration. In the first integration, we integrate the generalized fractional diffusion equation between the space coordinate x to the finite penetration depth δ. In the second integration, we integrate the generalized fractional diffusion equation between 0 to the finite penetration depth δ. We have the following results.

$$\bullet\bullet\bullet \int_0^\delta\int_x^\delta D_t^{\alpha,\rho}\bar{u}(x,t)dxdx = D_t^{\alpha,\rho}\int_0^\delta\int_x^\delta \left(1-\frac{x}{\delta}\right)^n dxdx$$
$$= D_t^{\alpha,\rho}\int_0^\delta \left[-\frac{\delta}{n+1}\left(1-\frac{x}{\delta}\right)^{n+1}\right]_x^\delta dx$$
$$= D_t^{\alpha,\rho}\int_0^\delta \frac{\delta}{n+1}\left(1-\frac{x}{\delta}\right)^{n+1} dx$$
$$= D_t^{\alpha,\rho}\left[-\frac{\delta^2}{(n+1)(n+2)}\left(1-\frac{x}{\delta}\right)^{n+1}\right]_0^\delta$$
$$= \frac{1}{(n+1)(n+2)}D_t^{\alpha,\rho}\delta^2. \tag{6.12}$$

$$\bullet\bullet\bullet \int_0^\delta\int_x^\delta \frac{\partial^2\bar{u}}{\partial x^2}dxdx = \int_0^\delta \left[\frac{\partial\bar{u}}{\partial x}\right]_x^\delta dx$$
$$= -\int_0^\delta \frac{\partial\bar{u}}{\partial x}dx$$
$$= \bar{u}(0,t).$$

6.4 Integral-Balance Methods

In this section, we describe the methods of finding an approximate solution of the generalized fractional diffusion equation defined by the following equation

$$D_t^{\alpha,\rho} u(x,t) = \mu \frac{\partial^2 u(x,t)}{\partial x^2}, \tag{6.13}$$

with initial boundary conditions defined as

- $u(x,0) = 0$ for $x > 0$,
- $u(0,t) = U_0 = 1$ for $t > 0$

We use the HBIM and the DIM [10, 11]. Note that the generalized fractional diffusion equations are parabolic equations. In this chapter, we assume that the profile of the generalized fractional diffusion equation is parabolic and defined with finite penetration depth δ. The problem consists of getting the penetration depth using the HBIM and the DIM.

6.4.1 Approximation with the HBIM

In this section, we develop the HBIM of finding the finite penetration depth δ. The method consists to integrate to both sides of the fractional diffusion equation (6.13) between 0 to δ. Applying the single integral between 0 to δ to both sides of equation (6.13) and assuming that the solution of the generalized fractional diffusion equation is defined by $\bar{u}(x,t) = \left(1 - \frac{x}{\delta}\right)^n$, we have the following relationships

$$\begin{aligned}
\int_0^\delta D_t^{\alpha,\rho} \bar{u}(x,t)dx &= \mu \int_0^\delta \frac{\partial^2 \bar{u}(x,t)}{\partial x^2} dx \\
&= \frac{\mu n}{\delta} \\
\frac{1}{n+1} D_t^{\alpha,\rho} \delta &= \frac{\mu n}{\delta} \\
D_t^{\alpha,\rho} \delta &= \frac{\mu n(n+1)}{\delta}.
\end{aligned} \tag{6.14}$$

Multiplying by the penetration depth δ to both sides of equation (6.14), we obtain the following fractional differential equation described by the Caputo-generalized fractional derivative

$$D_t^{\alpha,\rho} \delta^2 = 2\mu n(n+1). \tag{6.15}$$

Applying the ρ-Laplace transform [13] to both sides of equation (6.15), we obtain the following relationships

$$s^{\alpha}\bar{\delta}^2(s) = \frac{2\mu n(n+1)}{s}$$
$$\bar{\delta}^2(s) = \frac{2\mu n(n+1)}{s^{\alpha+1}}, \tag{6.16}$$

where $\bar{\delta}$ denotes the ρ-Laplace transform of δ. Applying the inverse of ρ-Laplace transform to both sides of equation (6.15), we obtain the following solution:

$$\delta^2(t) = \frac{2\mu n(n+1)t^{\alpha\rho}}{\rho^{\alpha}\sqrt{\Gamma(1+\alpha)}}$$
$$\delta(t) = \sqrt{2\mu n(n+1)}\frac{t^{\frac{\alpha\rho}{2}}}{\rho^{\frac{\alpha}{2}}\sqrt{\Gamma(\alpha+1)}}. \tag{6.17}$$

Finally, the penetration depth given by the HBIM is provided by

$$\delta(t) = \sqrt{2\mu n(n+1)}\frac{t^{\frac{\alpha\rho}{2}}}{\rho^{\frac{\alpha}{2}}\sqrt{\Gamma(\alpha+1)}} = \sqrt{\frac{2\mu n(n+1)}{\Gamma(1+\alpha)}}\left(\frac{t^{\rho}}{\rho}\right)^{\frac{\alpha}{2}}. \tag{6.18}$$

We observe that when we take the orders $\alpha = \rho = 1$, we recover the classical penetration depth obtained with the traditional diffusion model, see more information in [19]

$$\delta(t) = \sqrt{2\mu n(n+1)t}. \tag{6.19}$$

We observe that when the order $\rho = 1$, we recover the penetration depth obtained with the Caputo fractional derivative, for more information, see [9]

$$\delta(t) = \sqrt{2\mu n(n+1)}\frac{t^{\frac{\alpha}{2}}}{\sqrt{\Gamma(\alpha+1)}}. \tag{6.20}$$

We observe that when the order $\alpha = 1$, the penetration depth obtained with the Caputo-generalized fractional derivative is given by

$$\delta(t) = \sqrt{2\mu n(n+1)}\left(\frac{t^{\rho}}{\rho}\right)^{\frac{1}{2}}. \tag{6.21}$$

There exists a difference between the penetration depth in the classical diffusion equation and the penetration depth in the generalized fractional diffusion equation. In a fractional case, we have the fractional constant time defined by $\left(\frac{t^{\rho}}{\rho}\right)^{\frac{\alpha}{2}}$. It represents the fluctuation of the fractional order in the structure of the diffusion equation.

6.4.2 Approximation with DIM

In this section, we develop the DIM of getting the penetration depth δ. The method is described by Hristov in [9]. It consists of applying the double integration to both sides of the fractional diffusion equations. First, we integrate between x to δ. Second, we integrate between x to δ as is described in the following lines. Applying the double integration to both sides of equation (6.13) and assuming that $\bar{u}(x,t) = \left(1 - \frac{x}{\delta}\right)^n$ is the solution of the generalized fractional diffusion equation (6.13), we have the following

$$\begin{aligned}
\int_0^\delta \int_x^\delta D_t^{\alpha,\rho}\bar{u}(x,t)dx &= \mu \int_0^\delta \int_x^\delta \frac{\partial^2 \bar{u}(x,t)}{\partial x^2}dx \\
&= \mu u(0,t) \\
\frac{1}{(n+1)(n+2)} D_t^{\alpha,\rho}\delta^2 &= \mu \\
D_t^{\alpha,\rho}\delta^2 &= \mu(n+1)(n+2). \quad (6.22)
\end{aligned}$$

Applying the ρ-Laplace transform to both sides of equation (6.22), we get the following equations

$$\begin{aligned}
s^\alpha \bar{\delta}^2(s) &= \frac{\mu(n+1)(n+2)}{s} \\
\bar{\delta}^2(s) &= \frac{\mu(n+1)(n+2)}{s^{\alpha+1}}. \quad (6.23)
\end{aligned}$$

Applying the inverse of Laplace transform to both sides of equation (6.23), we get

$$\begin{aligned}
\delta^2(t) &= \frac{\mu(n+1)(n+2)t^{\alpha\rho}}{\rho^\alpha \Gamma(1+\alpha)} \\
\delta(t) &= \sqrt{\mu(n+1)(n+2)}\frac{t^{\frac{\alpha\rho}{2}}}{\rho^{\frac{\alpha}{2}}\sqrt{\Gamma(1+\alpha)}}. \quad (6.24)
\end{aligned}$$

Finally, the penetration depth with the DIM is given by

$$\delta(t) = \sqrt{\mu(n+1)(n+2)}\frac{t^{\frac{\alpha\rho}{2}}}{\rho^{\frac{\alpha}{2}}\sqrt{\Gamma(1+\alpha)}} = \sqrt{\frac{\mu(n+1)(n+2)}{\Gamma(1+\alpha)}}\left(\frac{t^\rho}{\rho}\right)^{\frac{\alpha}{2}}. \quad (6.25)$$

We observe that when the orders $\alpha = \rho = 1$, we recover the classical penetration depth obtained with the classical diffusion model (when we apply DIM)

$$\delta(t) = \sqrt{\mu(n+1)(n+2)t}. \quad (6.26)$$

We observe that when the order $\rho = 1$, we have the penetration depth obtained with the Caputo fractional derivative, see more information in [9]

$$\delta(t) = \sqrt{\mu(n+1)(n+2)}\frac{t^{\frac{\alpha}{2}}}{\sqrt{\Gamma(1+\alpha)}}. \quad (6.27)$$

We notice that when the order $\alpha = 1$, we have the penetration depth obtained with the Caputo-generalized fractional derivative given by

$$\delta(t) = \sqrt{\mu(n+1)(n+2)}\left(\frac{t^\rho}{\rho}\right)^{\frac{1}{2}}. \quad (6.28)$$

The time fractional constant $\left(\frac{t^\rho}{\rho}\right)^{\frac{\alpha}{2}}$ is the same in the two cases (HBIM and DIM). The penetration depths got with the HBIM and with the DIM depend on an exponent n. In the next section, we will discuss on this exponent n.

6.5 Approximate Solutions of the Generalized Fractional Diffusion Equations

In this section, we analyze two cases. We observe that the penetration depth got with the HBIM or the DIM depends on an exponent n. First, we assume that the profile of the fractional diffusion equation (6.13) described by the Caputo-generalized fractional derivative is quadratic, that is $n = 2$ [19]. Second, we assume that the profile of the fractional diffusion equation (6.13) described by the Caputo-generalized fractional derivative is cubic. That is $n = 3$ [19]. The assumed profile is defined by

$$\bar{u}(x,t) = \left(1 - \frac{x}{\delta}\right)^n. \tag{6.29}$$

which depends on the exponent n. The main question is what is the possible values of the exponent n. How to get this exponent n. In the literature, many works stipulated that the exponent n has integer values. They have supposed $n = 2$ or 3, see in [19]. The first exponent corresponds to the quadratic profile and the second exponent corresponds to the cubic profile. Another question concerns the method to get the exponent n. In [19], Myers and Mitchell propose a novel approach to get the exponent n. With their proposal, the authors find $n = 2.235$ for the HBIM and $n = 2,218$ for the refined integral method (RIM) (called here DIM) for the profile of the classical diffusion equation. For fractional diffusion equation with the Caputo fractional derivative and the Riemann–Liouville fractional derivative, many results for the possible values of the exponent n were proposed by Hristov [6–10, 12].

6.5.1 Quadratic Profile

In this section, we assume an approximate profile of the fractional diffusion equation described by equation (6.13), which is defined in the following form:

$$\bar{u}(x,t) = \left(1 - \frac{x}{\delta}\right)^2. \tag{6.30}$$

The objective is to give the complete form of the assumed profile of the fractional diffusion equation described by the Caputo-generalized fractional derivative. In the first case, we consider the penetration depth obtained with the HBIM. In the second case, we consider the penetration depth obtained with the DIM. We analyze the effect of the order ρ in the profile of the fractional diffusion equation described by the Caputo-generalized fractional derivative. For the rest, we fix $\alpha = 1$.

Approach with the HBIM: in this case, the profile of the fractional diffusion equation described by Caputo-generalized fractional derivative is given by

$$\bar{u}(x,t) = \left(1 - \frac{x}{\delta}\right)^n = \left(1 - \frac{x}{\sqrt{\frac{2\mu n(n+1)}{\Gamma(1+\alpha)}}\left(\frac{t^\rho}{\rho}\right)^{\frac{\alpha}{2}}}\right)^n.$$

With $n = 2$, we obtain the following approximate solution for the fractional diffusion equation described by the Caputo-generalized fractional derivative

$$\bar{u}(x,t) = \left(1 - \frac{x}{2\sqrt{3\mu}\left(\frac{t^\rho}{\rho}\right)^{\frac{1}{2}}}\right)^2. \tag{6.31}$$

The similarity variable plays an important role in the resolution of the fractional diffusion equation. The analytical solution of the fractional diffusion equation can be got using the similarity variable. The fractional similarity variable for the fractional diffusion equation described by the Caputo-generalized fractional derivative is given by $\eta = x/\sqrt{\mu\frac{t^{\rho\alpha}}{\rho}}$. It is no-Bolzmann similarity variable. In particular, when the order $\alpha = \rho = 1$, we recover the classical similarity variable given by $\eta = x/\sqrt{\mu t}$.

Let us analyze the effect of the order ρ in the diffusion processes. In Figure 6.1, we plot with $t = 0.9$, the profile of an approximate solution of the fractional diffusion equation when we fix the order $\alpha = 1$ and the values of the order ρ increase and less than 1. In this case, we recover the classical behaviour of diffusion equations. All the curves decay rapidly and converge to zero. Furthermore, the order of the profiles from left to right follows an increase in the order ρ. We notice an acceleration effect. The order ρ of the Caputo-generalized fractional derivative has an acceleration effect in the diffusion process.

In Figure 6.2, we plot with $t = 0.9$, the profile of an approximate solution of the fractional diffusion equation when we fix the order $\alpha = 1$, and the values of the order ρ decrease and are up to 1. We notice that all the curves are convex. It corresponds to the desired behaviour. The order of the profiles from right to left follows a decrease in the order ρ. We notice a retardation effect. The order ρ of the Caputo-generalized fractional derivative has a retardation effect in the diffusion process.

Conclusion: In general, the order ρ of the Caputo-generalized fractional derivative conserves the convex profile of the approximate solution of the

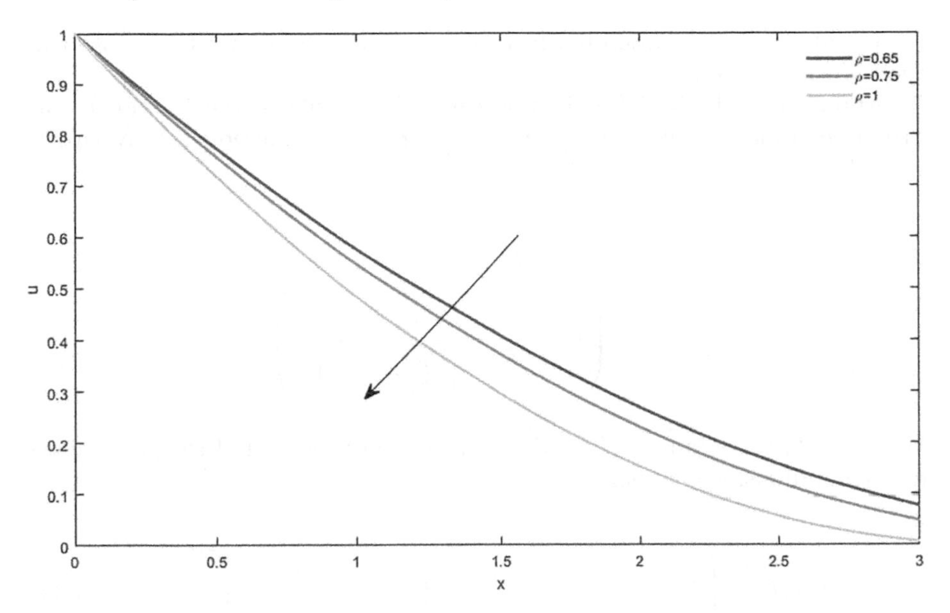

FIGURE 6.1
HBIM $\rho < 1$, $n = 2$.

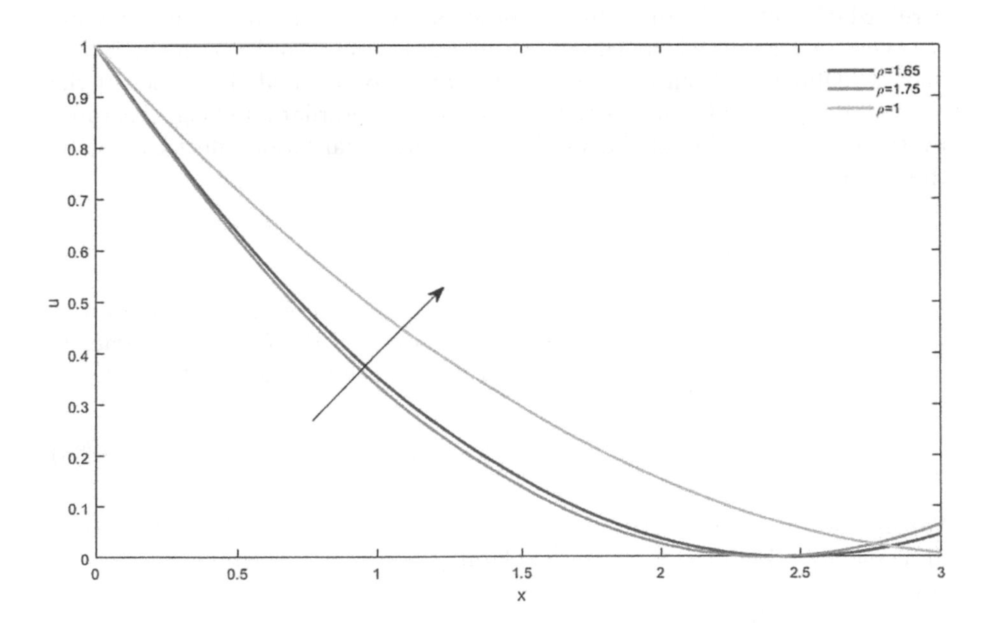

FIGURE 6.2
HBIM $\rho > 1$, $n = 2$.

fractional diffusion equation but has an acceleration or a retardation effect in the diffusion process.

Approach with the DIM: in this case, the profile of the fractional diffusion equation described by the Caputo-generalized fractional derivative is given by

$$\bar{u}(x,t) = \left(1 - \frac{x}{\delta}\right)^n$$
$$= \left(1 - \frac{x}{\sqrt{\frac{\mu(n+1)(n+2)}{\Gamma(1+\alpha)}}\left(\frac{t^\rho}{\rho}\right)^{\frac{\alpha}{2}}}\right)^n.$$

With $n = 2$, we obtain the following approximate solution of the generalized fractional diffusion equation

$$\bar{u}(x,t) = \left(1 - \frac{x}{2\sqrt{3\mu}\left(\frac{t^\rho}{\rho}\right)^{\frac{1}{2}}}\right)^2 = \left(1 - \frac{x}{\sqrt{\mu\frac{t^\rho}{\rho}}2\sqrt{3}}\right)^2. \tag{6.32}$$

The same conclusion as in the HBIM can be done. The order ρ of the Caputo-generalized fractional derivative conserves the convex profile of the approximate solution of the fractional diffusion equations but has an acceleration or a retardation effect in the diffusion process. We have no difference existing between the HBIM and the DIM when $n = 2$. In other words, both the HBIM and the DIM have been used. We fix the order $\alpha = 1$ and the values of the order ρ increase and is up to 1. We observe the retardation effect generated by the order ρ for the HBIM; we find the same retardation effect caused by the order ρ for DIM.

6.5.2 Cubic Profile

In this section, we stipulate an exponent n in the approximate solution of the fractional diffusion equation (6.13) described by the Caputo-generalized fractional derivative. The following form defines the assumed profile

$$\bar{u}(x,t) = \left(1 - \frac{x}{\delta}\right)^3. \tag{6.33}$$

We first express the solution of the approximate solution of the fractional diffusion equation described by the Caputo-generalized fractional derivative when the penetration depth is got by the HBIM. For the rest, we fix $\alpha = 1$.

Approach with the HBIM: in this case, the profile of the fractional diffusion equation described by the Caputo-generalized fractional derivative is given by

$$\bar{u}(x,t) = \left(1 - \frac{x}{\delta}\right)^n$$
$$= \left(1 - \frac{x}{\sqrt{\frac{2\mu n(n+1)}{\Gamma(1+\alpha)}} \left(\frac{t^\rho}{\rho}\right)^{\frac{\alpha}{2}}}\right)^n.$$

With $n = 3$, we obtain the following approximate solution of the generalized fractional diffusion equation

$$\bar{u}(x,t) = \left(1 - \frac{x}{2\sqrt{6\mu}\left(\frac{t^\rho}{\rho}\right)^{\frac{1}{2}}}\right)^3 = \left(1 - \frac{x}{\sqrt{\mu\frac{t^\rho}{\rho}}2\sqrt{6}}\right)^3. \tag{6.34}$$

Let us now analyze the effect of the order ρ in the diffusion process when we stipulate the exponent $n = 3$.

In Figures 6.3 and 6.4, we plot with $t = 0.9$, the profile of the approximate solution of the fractional diffusion equation. We fix the order $\alpha = 1$. The values of the order ρ increase and are less than 1. Clearly, all the curves decay rapidly and converge to zero. Furthermore, the order of the profiles from left to right follows a decrease in the order ρ. We notice an acceleration effect. The order ρ of the Caputo-generalized fractional derivative has an acceleration effect in the diffusion process.

In Figure 6.5, we plot with $t = 0.9$ the profile of the approximate solution of the generalized fractional diffusion equation. We fix the order $\alpha = 1$. The

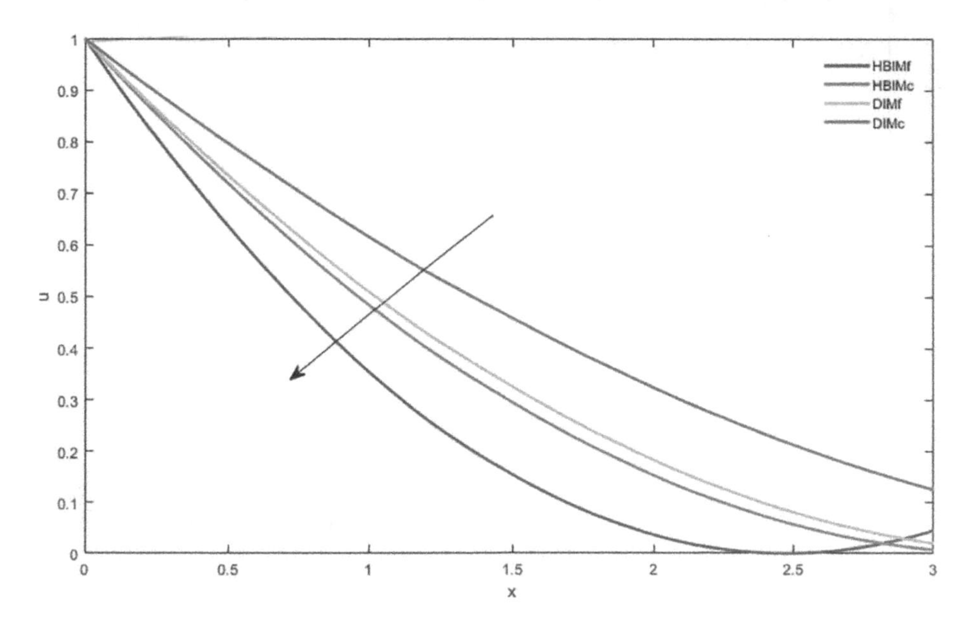

FIGURE 6.3

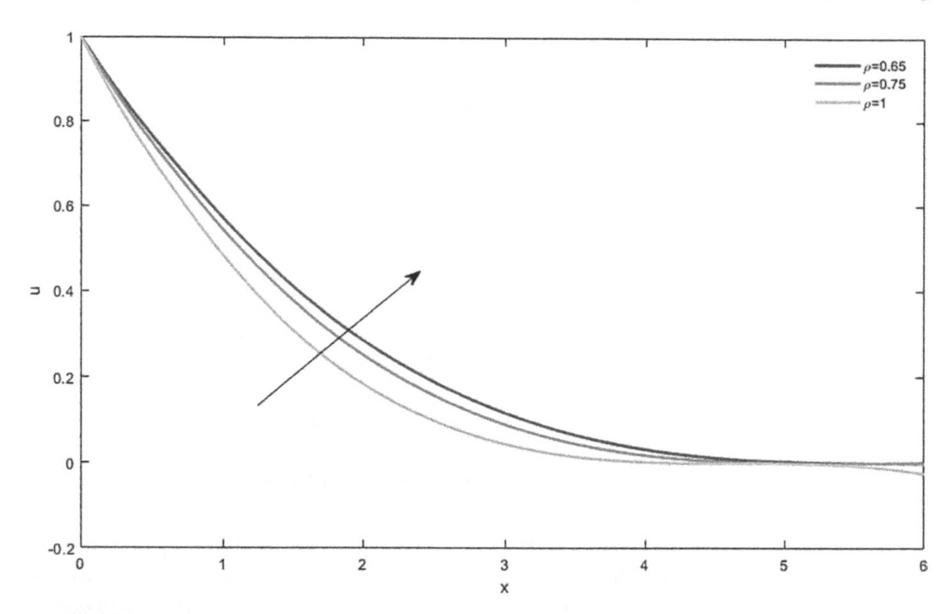

FIGURE 6.4
HBIM $\rho < 1$, $n = 3$.

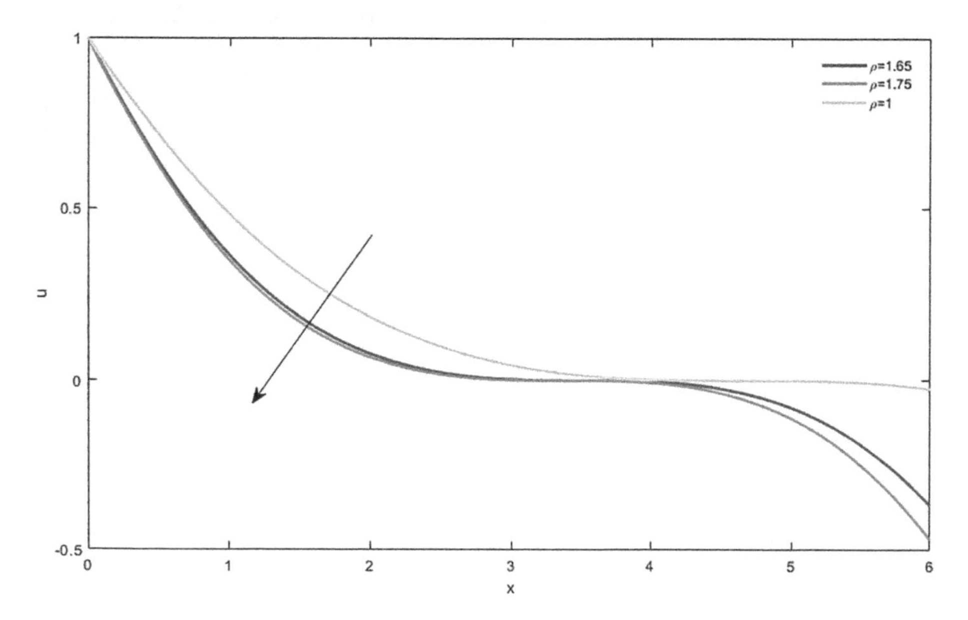

FIGURE 6.5
HBIM $\rho > 1$, $n = 3$.

values of the order ρ increase and is up to 1. We notice that all the curves decay. The order of the profiles from right to left follows an increase in the order ρ. We notice a retardation effect. The order ρ of the Caputo-generalized fractional derivative has a retardation effect in the diffusion process.

Let us now express the solution of an approximate solution of the fractional diffusion equation described by the Caputo-generalized fractional derivative when the penetration depth is got by the DIM.

Approach with the DIM: We stipulate the exponent $n = 3$. We obtain the following approximate solution of the generalized fractional diffusion equation

$$\bar{u}(x,t) = \left(1 - \frac{x}{2\sqrt{5\mu}\left(\frac{t^{\rho}}{\rho}\right)^{\frac{1}{2}}}\right)^{3} = \left(1 - \frac{x}{\sqrt{\mu\frac{t^{\rho}}{\rho}}2\sqrt{5\mu}}\right)^{3}. \tag{6.35}$$

Let us now compare the profile of the approximate solution obtained with the HBIM and that obtained with the DIM in some special cases. In Figure 6.6 are depicted the profiles of the approximate solutions with HBIM and DIM. The difference between the DIM and the HBIM is: with the HBIM, all the curves decrease and converge to zero, but the diffusion process is globally more rapid than the diffusion process in the DIM, when the order ρ increase and is up to 1. In other words, with DIM, the process of diffusion is slow. When the order ρ increases and is less than 1, with the DIM all the curves decrease

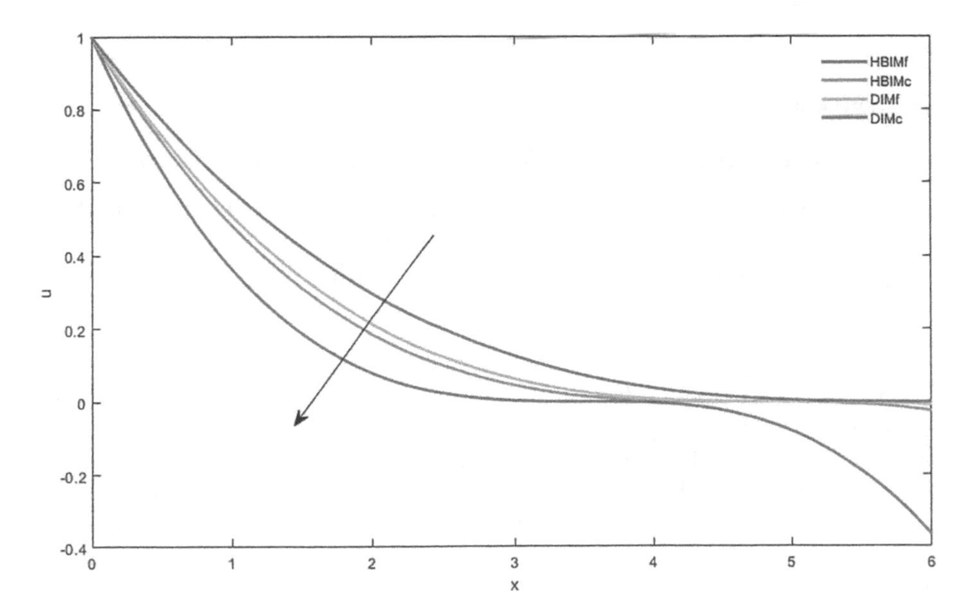

FIGURE 6.6
HBIM and DIM$\rho > 1$,$n = 3$.

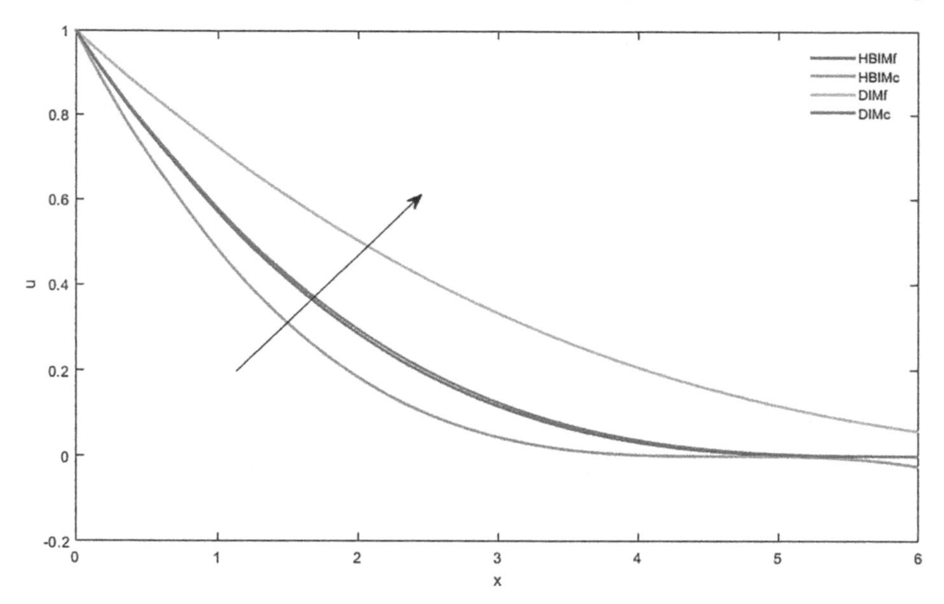

FIGURE 6.7
HBIM and DIM $\rho < 1$, $n = 3$.

and converge to zero, but the diffusion process is globally more rapid than the diffusion process in the HBIM. In other words, with the HBIM, the process of diffusion is slow (Figure 6.7).

6.6 Myers and Mitchell Approach for Exponent n

In this section, we discuss on the exponent n. In many contributions, the authors use the exponent $n = 2$ or $n = 3$, see in [19]. We observe that when $n = 2$ we obtain a second-order polynomial for which the discriminant is null, that is

$$\Delta = \frac{4}{\delta^2} - \frac{4}{\delta^2} = 0.$$

It corresponds to the profile of a parabolic equation. There exist in the literature, a method proposed by Myers and Mitchell in [19] of getting the exponent n. This method was used in many papers related to integral-balance method proposed by Hristov [6–10, 12]. The technique proposed by Myers and Mitchell of getting the exponent n consists of minimizing the square of the residual function introduced by Langford in integer order. The Langford function is defined in the following form

$$\int_0^1 \xi^2(\bar{u}(x,t))dx = \int_0^1 \left[D_t^{\alpha,\rho}\bar{u}(x,t) - \mu \frac{\partial^2 \bar{u}(x,t)}{\partial x^2} \right]^2 dx, \tag{6.36}$$

where the residual function is given by

$$\xi(\bar{u}(x,t)) = D_t^{\alpha,\rho}\bar{u}(x,t) - \mu\frac{\partial^2\bar{u}(x,t)}{\partial x^2}. \tag{6.37}$$

Langford has proposed the following criterion: the function $\int_0^1 \xi^2(\bar{u}(x,t))dx$ attend it minimum when the exponent n is optimal. In other words, when $\bar{u}$ approach u, the function $\xi(\bar{u}(x,t))$ should approach it minimum over the entire penetration depth [6–10, 12]. Let us now describe the resolution of the minimization problem in the context of non-integer order, and furthermore, when the used fractional derivative is the Caputo-generalized fractional derivative.

6.6.1 Residual Function

In this subsection, we establish the value of the residual function. We have to evaluate the fractional derivative of the assumed profile. We have the following

$$D_t^{\alpha,\rho}\bar{u}(x,t)dx = D_t^{\alpha,\rho}\left(1-\frac{x}{\delta}\right)^n.$$

Note at zero you can use the Taylor expansion of the assumed function, and we get

$$\begin{aligned}
D_t^{\alpha,\rho}\bar{u}(x,t)dx &= D_t^{\alpha,\rho}\left(1-\frac{x}{\delta}\right)^n \\
&= D_t^{\alpha,\rho}\left[1-\frac{nx}{\delta}+\frac{n(n+1)x^2}{2\delta^2}\right] \\
&= -nxD_t^{\alpha,\rho}\delta^{-1}+\frac{n(n+1)x^2}{2}D_t^{\alpha,\rho}\delta^{-2}.
\end{aligned} \tag{6.38}$$

The second part of the calculation of the residual function is to evaluate the second-order derivative of the assumed profile, we have

$$\frac{\partial^2\bar{u}(x,t)}{\partial x^2} = \frac{n(n-1)}{\delta^2}\left(1-\frac{x}{\delta}\right)^{n-2}. \tag{6.39}$$

Combining equation (6.38) and (6.39), we get the residual function defined by

$$\xi(\bar{u}(x,t)) = -nxD_t^{\alpha,\rho}\delta^{-1}+\frac{n(n+1)x^2}{2}D_t^{\alpha,\rho}\delta^{-2}-\frac{n(n-1)}{\delta^2}\left(1-\frac{x}{\delta}\right)^{n-2}. \tag{6.40}$$

Then, the residual function using the HBIM penetration depth is in the following form

$$\begin{aligned}
\xi(\bar{u}(x,t)) = &-\frac{nx}{\sqrt{2\mu n(n+1)}}D_t^{\alpha,\rho}\left(\frac{t^\rho}{\rho}\right)^{-\frac{\alpha}{2}}+\frac{(n-1)x^2}{4\mu(n+1)}D_t^{\alpha,\rho}\left(\frac{t^\rho}{\rho}\right)^{-\alpha} \\
&-\frac{(n-1)}{2\mu(n+1)}\left(\frac{t^\rho}{\rho}\right)^{-\alpha}\left(1-\frac{x}{\sqrt{2\mu n(n+1)}}\left(\frac{t^\rho}{\rho}\right)^{-\frac{\alpha}{2}}\right)^{n-2}.
\end{aligned} \tag{6.41}$$

The residual function using the DIM and the penetration depth is in the following form:

$$\xi(\bar{u}(x,t)) = -\frac{nx}{\sqrt{\mu(n+1)(n+2)}} D_t^{\alpha,\rho}\left(\frac{t^\rho}{\rho}\right)^{-\frac{\alpha}{2}} + \frac{n(n-1)x^2}{2\mu(n+1)(n+2)} D_t^{\alpha,\rho}\left(\frac{t^\rho}{\rho}\right)^{-\alpha} - \frac{n(n-1)}{\mu(n+1)(n+2)}\left(\frac{t^\rho}{\rho}\right)^{-\alpha}\left(1 - \frac{x}{\sqrt{\mu(n+1)(n+2)}}\left(\frac{t^\rho}{\rho}\right)^{-\frac{\alpha}{2}}\right)^{n-2}. \tag{6.42}$$

In addition, we recall the Caputo-generalized fractional derivative of the function $\left(\frac{t^\rho}{\rho}\right)^\beta$, that is

$$D^{\alpha,\rho}\left(\frac{t^\rho}{\rho}\right)^\beta = \frac{\Gamma(\beta+1)}{\Gamma(\beta-\alpha+1)}\left(\frac{t^\rho}{\rho}\right)^{\beta-\alpha}. \tag{6.43}$$

6.6.2 At Boundary Conditions

We investigate of getting the exponent n at the boundary conditions. Note that the exponent n at the boundary conditions is not necessarily an optimal exponent n. In this section, we will not use Langford criterion of getting n. The contribution of this section is to verify the Goodman conditions at the boundary conditions. That is

$$\bar{u}(\delta,t) = \frac{\partial \bar{u}(\delta,t)}{\partial x} = 0. \tag{6.44}$$

Note that when $\bar{u}$ is a solution of the fractional diffusion equation described by the Caputo-generalized fractional derivative, then the residual function is null, that is $\xi(\bar{u}(x,t)) = 0$.

At the point $x = 0$, the residual function using equation (6.41) for the HBIM or equation (6.42) for the DIM is given by

$$\xi(\bar{u}(0,t)) = \frac{n(n-1)}{\delta^2} = 0. \tag{6.45}$$

From which we easily get the exponent at the first boundary condition $n = 1$. To satisfy the Goodman boundary conditions, we must impose that the residual function at $x = 0$ must be strictly positive. In other words, $\xi(\bar{u}(0,t)) > 0$ the exponent n must satisfy $n > 1$.

At $x = \delta$ or when the coordinate x approaches the penetration depth δ, to recover Goodman boundary conditions, we must impose the residual function when x approaches δ must be strictly positive. That is using equation (6.41) for the HBIM or equation (6.42) for the DIM

$$\lim_{x\to\delta} \xi(\bar{u}(\delta,t)) = \lim_{x\to\delta} \frac{n(n-1)}{\delta^2}\left(1 - \frac{x}{\delta}\right)^{n-2} > 0. \tag{6.46}$$

Thus, the exponent n must satisfy $n > 2$. As it is said in the previous section, the exponent $n = 1$ and $n = 2$ got at the boundary conditions do not guarantee an optimal exponent n satisfying Langford criterion.

6.6.3 Outsides of Boundary Conditions

In this section, we give the procedure of finding the exponent n when $x > 0$ and $x \neq \delta$. To get the exponent n according to Langford's criterion, we use the Myers and Mitchell method in integer order. We minimize the square of the residual function proposed by Langford

$$\int_0^1 \xi^2(\bar{u}(x,t))dx = \int_0^1 \left[D_t^{\alpha,\rho}\bar{u}(x,t) - \mu \frac{\partial^2 \bar{u}(x,t)}{\partial x^2} \right]^2 dx, \tag{6.47}$$

where the residual function is given by

$$\xi(\bar{u}(x,t)) = -nxD_t^{\alpha,\rho}\delta^{-1} + \frac{n(n+1)x^2}{2} D_t^{\alpha,\rho}\delta^{-2} - \frac{n(n-1)}{\delta^2}\left(1 - \frac{x}{\delta}\right)^{n-2}. \tag{6.48}$$

Finally, we have to get an optimal exponent n under which the function

$$\begin{aligned}\int_0^1 \xi^2(\bar{u}(x,t))dx = \int_0^1 \Bigg[& -nxD_t^{\alpha,\rho}\delta^{-1} + \frac{n(n+1)x^2}{2} D_t^{\alpha,\rho}\delta^{-2} \\ & - \frac{n(n-1)}{\delta^2}\left(1 - \frac{x}{\delta}\right)^{n-2}\Bigg]^2 dx,\end{aligned} \tag{6.49}$$

is minimum. Hristov proposed many works corresponding to the exponent n in the context of fractional order derivative operators (the Caputo and the Riemann–Liouville fractional derivative) in [6–10, 12]. Note that Myers and Mitchell get when $\alpha = \rho = 1$ an optimal exponent $n = 2.235$ for the HBIM and $n = 2.218$ for the DIM. For simplification in this chapter, we accept Hristov exponent in the non-integer order time derivative and Myers and Mitchell exponent in the integer order time derivative. It is proved in the literature, in general, the possible values of the exponent n are into the interval $[2, 3]$.

Finding a more accurate method of getting an optimal exponent n in integer order context or non-integer order is a paradox, and we hope the future works will answer this open problem. Myers and Mitchell begin to answer this problem, but the proposed method is restricted to the non-linear subdiffusion equation. Myers and Mitchell's second method described in [6–10,12] proposes the penetration depth obtained with HBIM and DIM are equivalents. This method can be applied here with the generalized fractional diffusion equations. Applying this method here, we get

$$\begin{aligned}\sqrt{2\mu n(n+1)}\left(\frac{t^\rho}{\rho}\right)^{\frac{\alpha}{2}} &= \sqrt{\mu(n+1)(n+2)}\left(\frac{t^\rho}{\rho}\right)^{\frac{\alpha}{2}} \\ 2n &= n+2. \\ n &= 2.\end{aligned} \tag{6.50}$$

We notice that the resolution of equation (6.50) is possible here. We recover a quadratic profile. Conclusion: for the generalized fractional diffusion equation, the method is adequate.

6.7 Conclusion

In this chapter, we have discussed the integral-balance methods (the HBIM and the DIM) to approach the profile of the fractional diffusion equation described by the Caputo-generalized fractional derivative. We have found that the order ρ of the Caputo-generalized fractional derivative has a retardation or an acceleration effect in the diffusion process. We have also discussed in the method to find an exponent n in the profile of the approximate solution and open new directions of investigation related to the exponent n. Finding a more simple method of getting an optimal exponent n in integer order context or non-integer order context is a paradox, and we hope in future works, this paradox will be solved.

References

[1] Atangana, A. and Baleanu, D., New fractional derivatives with nonlocal and non-singular kernel: Theory and application to heat transfer model, arXiv preprint arXiv:1602.03408, (2016).

[2] Caputo, M. and Fabrizio, M., A new definition of fractional derivative without singular kernel, *Progress in Fractional Differentiation and Applications*, **1(2)**, 1–15, (2015).

[3] Hristov, J., Approximate solutions to fractional subdiffusion equations, *The European Physical Journal Special Topics*, **193(1)**, 229–243, (2011).

[4] Hristov, J., Transient heat diffusion with a non-singular fading memory: From the cattaneo constitutive equation with Jeffrey's kernel to the Caputo-Fabrizio time-fractional derivative, *Thermal Science*, **20(2)**, 757–762, (2016).

[5] Hristov, J., Multiple integral-balance method basic idea and an example with mullin's model of thermal grooving, *Thermal Science*, **2(1)**, 555–1560, (2017).

[6] Hristov, J., The non-linear dodson diffusion equation: Approximate solutions and beyond with formalistic fractionalization, *Mathematics in Natural Science*, **1(1)**, 1–17, (2017).

[7] Hristov, J., Space-fractional diffusion with a potential power-law coefficient: Transient approximate solution, *Progress in Fractional Differentiation and Applications*, **3(1)**, 19–39, (2017).

[8] Hristov, J., Fourth-order fractional diffusion model of thermal grooving: Integral approach to approximate closed form solution of the Mullins model, *Mathematical Modelling of Natural Phenomena*, **13**, 6, (2018).

[9] Hristov, J., Integral-balance solution to nonlinear subdiffusion equation, *Frontiers in Fractional Calculus*, **1**, 70, (2018).

[10] Hristov, J., The heat radiation diffusion equation: Explicit analytical solutions by improved integral-balance method, *Thermal Science*, **22(2)**, 777–778, (2018).

[11] Hristov, J., Integral balance approach to 1-d space-fractional diffusion models, *Mathematical Methods in Engineering*, 111–131, Springer.

[12] Hristov, J., A transient flow of a non-newtonian fluid modelled by a mixed time-space derivative: An improved integral-balance approach, *Mathematical Methods in Engineering*, 153–174, (2019), Springer.

[13] Fahd, J. and Abdeljawad, T., A modified Laplace transform for certain generalized fractional operators, *Results in Nonlinear Analysis*, **2**, 88–98, (2018).

[14] Fahd, J., Abdeljawad, T., and Baleanu, D., On the generalized fractional derivatives and their Caputo modification, *Journal of Nonlinear Sciences and Applications*, **10**, 2607–2619, (2017).

[15] Katugampola U. N., Theory and applications of fractional differential equations, *Applied Mathematics and Computation*, **218(3)**, 860–865, (2011).

[16] Kilbas, A. A., Srivastava, H. M., and Trujillo, J. J., *Theory and Applications of Fractional Differential Equations*, Elsevier, (2006).

[17] Mitchell, S. L., Applying the combined integral method to two-phase stefan problems with delayed on set of phase change, *Journal of Computational and Applied Mathematics*, **281**, 58–73, (2015).

[18] Mitchell, S. L., and Myers, T. G., Heat balance integral method for one-dimensional finite ablation, *Journal of Thermophysics and Heat Transfer*, **22(3)**, 508–514, (2008).

[19] Myers, T. G., Optimal exponent heat balance and refined integral methods applied to Stefan problems, *International Journal of Heat and Mass Transfer*, **53**, 1119–1127, (2010).

[20] Sene, N., On stability analysis of the fractional nonlinear systems with Hurwitz state matrix, *Journal of Fractional Calculus and Applied Analysis*, **10**, 1–9, (2019).

[21] Sene, N., Exponential form for Lyapunov function and stability analysis of the fractional differential equations, *Journal of Mathematical and Computational Science*, **18(4)**, 388–397, (2018).

[22] Sene, N., Lyapunov characterization of the fractional nonlinear systems with exogenous input, *Fractal and Fractional*, **2(2)**, (2018).

[23] Sene, N., Solutions for some conformable differential equations, *Progress in Fractional Differentiation and Applications*, **4(4)**, 493–501, (2018).

[24] Sene, N., Stokes' first problem for heated flat plate with Atangana–Baleanu fractional derivative, *Chaos, Solitons & Fractals*, **117**, 68–75, (2018).

[25] Sheikh, N. A., Ali, F., Saqib, M., Khan, I., Jan, S. A. A., Alshomrani, A. S., and Alghamdi, M. S., Comparison and analysis of the Atangana–Baleanu and Caputo–Fabrizio fractional derivatives for generalized Casson fluid model with heat generation and chemical reaction, *Results in Physics*, **7**, 789–800, (2017).

7

A Hybrid Formulation for Fractional Model of Toda Lattice Equations

Amit Kumar
Balarampur College

Sunil Kumar and Ranbir Kumar
National Institute of Technology

CONTENTS

7.1 Introduction

In the past few decades, considerable interests in fractional differential equations have been simulated due to their numerous applications in the area of physics and engineering [1]. Many important phenomena are well described by fractional differential equation in electromagnetic, acoustics, viscoelasticity, material science, electrochemistry, diffusion, control, relaxation processes and so on [2–10]. Finding efficient and accurate methods for solving fractional differential equations has been an active research in this time. It is not easy to find exact solutions of the non-linear fractional differential equations arising in science and engineering; hence, analytical and numerical methods must be used.

One new hybrid method yielding series solutions is called the homotopy analysis transform method (HATM). Actually, the homotopy analysis method (HAM) was first introduced and applied by Liao [11–17] in 1992. The HAM has been successfully applied by many researchers for solving linear and non-linear partial differential equations [18–25, 33–36]. The basic motivation of the present work is to develop a hybrid algorithm to handle the fractional Toda lattice equations. The proposed hybrid method is amagalation of the HAM, Laplace transform method and Adomian's polynomials. The main advantage of this proposed method is its capability of combining two powerful methods with Adomian's polynomials for obtaining rapid convergent series for fractional partial differential equations. The HATM also allows us to effectively control the region of convergence and rate of convergence of a series solution to a non-linear partial differential equation via control of an initial approximation, an auxiliary linear operator, an auxiliary function and a convergence control parameter. The HATM solution generally agrees with the exact solution at large domains when compared with homotopy perturbation method and HAM solutions.

In this chapter, we intend to extend the application of the proposed method to solve the fractional order Toda lattice equations. Recently, many researchers [26–30] have applied to obtain the proficient solutions of the many differential and integral equations by coupling HAM and Laplace transform method. Recently, Mokhtari [31] has solved the Toda lattice equations by using variational iteration method. However, the following Toda lattice equations with fractional order have not yet been solved by any researchers. The beauty of this chapter is the introduction of new type of residual error [18, 37–38] which helps to find out the optimal values of auxiliary parameter for getting better convergence of the solution. The fractional order Toda lattice equations have numerous applications because of their use in physical and mathematical models of enormous real-world phenomena like biophysical systems, atomic chains with on-site cubic non-linearities, electrical lattices, molecular crystals and recently in arrays of coupled non-linear optical wave guides. For more application of the methods, see [39–46].

The chapter is organized as follows. We begin by introducing some basic definitions of Laplace transform which are required for establishing our results. In Section 7.2, the basic idea of the new HATM for fractional partial differential equations is presented. Section 7.3 is devoted to applying the new HATM for time-fractional Toda lattice equations. Numerical result and discussion are reported in Section 7.4. In Section 7.5, we give our conclusions briefly.

Definition 7.1 The Laplace transform of the function $f(t)$ is defined by

$$F(s) = L[f(t)] = \int_0^\infty e^{-st} f(t)\mathrm{d}t. \tag{7.1}$$

Definition 7.2 The Laplace transform $L[u(x,t)]$ of the Riemann–Liouville fractional integral is defined as [4]:

$$L[I_t^\alpha u(x,t)] = s^{-\alpha} L[u(x,t)]. \tag{7.2}$$

Definition 7.3 The Laplace transform $L[u(x,t)]$ of the Caputo-fractional derivative is defined as [4]:

$$L\left[D_t^{n\alpha}u(x,t)\right] = s^{n\alpha}L[u(x,t)] - \sum_{k=0}^{n-1} s^{(n\alpha-k-1)}u^{(k)}(x,0), \quad n-1 < n\alpha \leq n. \tag{7.3}$$

7.2 Basic Idea of HATM with Adomian's Polynomials

To illustrate the basic idea of the HATM for the fractional partial differential equation arising in science and engineering, we consider the following fractional partial differential equation:

$$\begin{aligned} &D_t^{n\alpha}u_j(x,t) + R_j[x]u_j(x,t) + N_j[x]u_j(x,t) = g_j(x,t), \\ &\quad t > 0, x \in R,\ j = 1,2,\dots k,\ n-1 < n\alpha \leq n, \end{aligned} \tag{7.4}$$

where $D_t^{n\alpha} = \frac{\partial^{n\alpha}}{\partial t^{n\alpha}}$, is the Caputo-fractional derivative $R_j[x]$ is the linear operator in x, $N_j[x]$, $j = 1,2,3\dots,k$ is the general non-linear operator in x and $g_j(x,t), j = 1,2,3\dots,k$ are continuous functions. For simplicity, we ignore all initial and boundary conditions, which can be treated in a similar way. Now the methodology consists of applying the Laplace transform first on both sides of Eq. (7.4), and thus we get

$$\begin{aligned} &L\left[D_t^{n\alpha}u_j(x,t)\right] + L[R_j[x]u_j(x,t) + N_j[x]u_j(x,t)] = L\left[g_j(x,t)\right], \\ &\quad t > 0, x \in R,\ j = 1,2,\dots k,\ n-1 < n\alpha \leq n. \end{aligned} \tag{7.5}$$

Now, using the differentiation property of the Laplace transform, we have

$$\begin{aligned} L\left[u_j(x,t)\right] &- \frac{1}{s^{n\alpha}}\sum_{k=0}^{n-1} s^{(n\alpha-k-1)}u_j^k(x,0) + \frac{1}{s^{n\alpha}}L\left(R_j[x]u_j(x,t)\right. \\ &\left. + N_j[x]u_j(x,t) - g_j(x,t)\right) = 0. \quad t > 0, x \in R,\ j = 1,2,\dots k, \\ &\qquad n-1 < n\alpha \leq n. \end{aligned} \tag{7.6}$$

We define the non-linear operator as

$$\begin{aligned} \eta_j\left[\phi_j(r,t;q)\right] = L\left[\phi_j(x,t;q)\right] &- \frac{1}{s^{n\alpha}}\sum_{k=0}^{n-1} s^{(n\alpha-k-1)}u_j^k(x,0) \\ &+ \frac{1}{s^{n\alpha}}L\left(R_j[x]u_j(x,t) + N_j[x]u_j(x,t) - g_j(x,t)\right), \\ &t > 0, x \in R,\ j = 1,2,3\dots k,\ n-1 < n\alpha \leq n. \end{aligned} \tag{7.7}$$

where $q \in [0,1]$ be an embedding parameter and $\phi_j(x,t;q)$ is the real function of x,t and q. By means of generalizing traditional homotopy methods, Liao [11–17] constructed the zero-order deformation equation

$$(1-q)L\left[\phi_j(x,t;q)-u_{j,0}(x,t)\right]=\hbar qH(x,t)\eta_j\left[\phi_j(x,t:q)\right],$$
$$t>0, x\in R,\quad j=1,2,3\ldots k,\ n-1<n\alpha\leq n. \tag{7.8}$$

where $\hbar$ is a non-zero auxiliary parameter, $H(x,t)\neq 0$ is an auxiliary function, $u_{0,j}(x,t)$ is an initial guess of $u_j(x,t)$ and $\phi_j(x,t;q)$ is an unknown function. It is important that one has great freedom to choose auxiliary thing in HATM. Obviously, when $q=0$ and $q=1$, it holds

$$\phi_j(x,t;0)=u_{j,0}(x,t),\quad \phi_j(x,t;1)=u_j(x,t), \tag{7.9}$$

respectively. Thus, as q increases from 0 to 1, the solution varies from the initial guess $u_{j,0}(x,t)$ to the solution $u_j(x,t)$. Expanding $\phi_j(x,t;q)$ in Taylor's series with respect to q, we have

$$\phi_j(x,t;q)=u_{j,0}(x,t)+\sum_{m=1}^{\infty}q^m u_{j,m}(x,t),\quad j=1,2,\ldots k. \tag{7.10}$$

where

$$u_{j,m}(x,t)=\frac{1}{m!}\left.\frac{\partial^m\phi_j(x,t;q)}{\partial q^m}\right|_{q=0}. \tag{7.11}$$

The convergence of series solution (7.10) is controlled by $\hbar$. If the auxiliary linear operator, the initial guess, the auxiliary parameter $\hbar$ and the auxiliary function are properly chosen, the series (7.10) converges at $q=1$, and thus we have

$$u_j(x,t)=u_{j,0}(x,t)+\sum_{m=1}^{\infty}u_{j,m}(x,t),\ j=1,2,\ldots k. \tag{7.12}$$

which must be one of the solutions of original non-linear equations. The above expression provides us with a relationship between the initial guess $u_{j,0}(x,t)$ and the exact solution $u_j(x,t)$ by means of the terms $u_{j,m}(x,t)$ that are still to be determined.

We define the following vector

$$\vec{u}_{j,n}=\{u_{j,0}(x,t),u_{j,1}(x,t),\ u_{j,2}(x,t),\ldots,u_{j,n}(x,t)\}. \tag{7.13}$$

Differentiating the zero-order deformation Eq. (7.8) m time with respect to embedding parameter q and then setting $q=0$ and finally dividing them by $m!$, we obtain the mth-order deformation equation

$$L\left[u_{j,m}(x,t)-\chi_m u_{j,m-1}(x,t)\right]=\hbar qH(x,t)R_{j,m}\left(\vec{u}_{j,m-1},x,t\right). \tag{7.14}$$

Operating the inverse Laplace transform on both sides, we get

$$u_{j,m}(x,t)=\chi_m u_{j,m-1}(x,t)+\hbar qL^{-1}\left[H(x,t)R_{j,m}\left(\vec{u}_{j,m-1},x,t\right)\right], \tag{7.15}$$

where

$$R_{j,m}\left(\vec{u}_{j,m-1},x,t\right)=\frac{1}{(m-1)!}\left.\frac{\partial^{m-1}\phi_j(x,t;q)}{\partial q^{m-1}}\right|_{q=0}, \tag{7.16}$$

and

$$\chi_m = \begin{cases} 0, m \leq 1, \\ 1, m > 1. \end{cases}$$

In this way, it is easy to obtain $u_{j,m}(x,t)$ for $m \geq 1$ at Mth order, and we have

$$u_j(x,t) = \sum_{m=0}^{M} u_{j,m}(x,t), \quad j = 1, 2, \ldots k. \tag{7.17}$$

when $M \to \infty$, we get an accurate approximation of the original Eq. (7.4).

$$u_{n,m}(t) = (\chi_m + \hbar)u_{n,m-1} - \hbar(1 - \chi_m)L^{-1}\left(\sum_{k=0}^{j-1} \frac{u^k(0)}{s^{k+1}}\right)$$
$$+ \hbar L^{-1}\left(\frac{1}{s^{j\alpha}} L\left(R_n[t]u_n(t) + N_n[t]u_n(t) - g_n(t)\right)\right). \tag{7.18}$$

In this way, it is easy to obtain $u_{n,\,m}(t)$ for $m \geq 1$ at Mth order, and we have

$$u_n(t) = u_{n,0}(t) + \lim_{k\to\infty} \sum_{m=1}^{k} u_{n,m}(t), \tag{7.19}$$

where $u_{n,0}(t) = u_n(0) + tu'_n(0) + t^2u''_n(0) + t^3u'''_n(0) + \cdots + t^{j-1}u^{(j-1)}{}_n(0) = \sum_{i=0}^{j-1} t^i u_n^{(i-1)}(0)$.

The non-linear term $N_n[t]u_n(t)$ is expanded in terms of Adomian's polynomials as

$$N_n[t]u_n(t) = N_n\left(\sum_{m=0}^{k} u_{n,m}(t)\right) = \sum_{m=0}^{k} A_{n,m}(u_{n,0},\, u_{n,1},\, u_{n,2},\, u_{n,3}, \ldots, u_{n,m}), \tag{7.20}$$

where $A_{n,m}$ are the Adomian's polynomials that are calculated by algorithm (7.20) constructed by Adomian [32]:

$$A_{n,k}(u_{n,0},\, u_{n,1},\, u_{n,2},\, u_{n,3}, \ldots, u_{n,k}) = \frac{1}{n!}\left[\frac{d^k}{dt^k} N_n\left(\sum_{p=0}^{k} \lambda^p u_{n,p}\right)\right]_{\lambda=0}, \quad k \geq 0. \tag{7.21}$$

Combining Eqs. (7.17) and (7.20), we get

$$u_{n,m}(t) = (\chi_m + \hbar)u_{n,m-1} - \hbar(1 - \chi_m)\sum_{i=0}^{j-1} t^i u_n^{(i-1)}(0)$$
$$+ \hbar L^{-1}\left(\frac{1}{s^{j\alpha}} L\left(R_{n,m-1}[t]u_{n,m-1}(t)\right.\right.$$
$$\left.\left. + \sum_{k=0}^{m-1} A_{n,k}(u_{n,0},\, u_{n,1},\, u_{n,2},\, u_{n,3}, \ldots, u_{n,m}) - g_n(t)\right)\right).$$
$$m = 1, 2, 3, \ldots \tag{7.22}$$

From Eq. (7.21), we calculate the various $u_{n,m}(t)$ for $n \geq 1$, and substituting these values in (7.18), we obtain the analytical approximate solution of Eq. (7.4). The novelty of our proposed algorithm is that a new correction functional (7.22) is constructed by expanding the non-linear term as a series of Adomian's polynomials in Eq. (7.22). We combine the new hybrid iterative algorithm for non-linear fractional partial differential equations arising in different branches of science and engineering.

7.3 Application to the Toda Lattice Equations

In this section, we discuss the implementation of our new modified method to the following Toda lattice equation [31]:

$$\begin{cases} D_t^\alpha u_n = u_n\left(v_n - v_{n-1}\right), \\ D_t^\alpha v_n = v_n\left(u_{n+1} - u_n\right), \end{cases} \qquad t \geq 0, 0 < \alpha \leq 1, \tag{7.23}$$

with initial conditions

$$\begin{aligned} u_0(n,t) &= u(n,0) = -c\coth(d) + c\tanh(dn), \\ v_0(n,t) &= v(n,0) = -c\coth(d) - c\tanh(dn). \end{aligned} \tag{7.24}$$

Applying the Laplace transform on both sides in Eq. (7.23) and after using the differentiation property of Laplace transform of fractional derivative, we get

$$\begin{aligned} s^\alpha L\left[u_n\left(n,\ t\right)\right] - s^{\alpha-1} u_n\left(n,0\right) &= L\left[u_n\left(v_n - v_{n-1}\right)\right], \\ s^\alpha L\left[v_n\left(n,\ t\right)\right] - s^{\alpha-1} v_n\left(n,0\right) &= L\left[v_n\left(u_{n+1} - u_n\right)\right]. \end{aligned} \tag{7.25}$$

On simplifying, we get

$$\begin{aligned} L\left[u_n\left(n,t\right)\right] &= \frac{1}{s}\left(-c\coth\left(d\right) + c\tanh\left(dn\right)\right) + \frac{1}{s^\alpha} L\left[u_n\left(v_n - v_{n-1}\right)\right]. \\ L\left[v_n\left(n,t\right)\right] &= \frac{1}{s}\left(-c\coth\left(d\right) - c\tanh\left(dn\right)\right) + \frac{1}{s^\alpha} L\left[v_n\left(u_{n+1} - u_n\right)\right]. \end{aligned} \tag{7.26}$$

We choose the linear operator as

$$\pounds\left[\phi_j\left(n,t;q\right)\right] = L\left[\phi_j\left(n,t;q\right)\right], \quad j = 1, 2. \tag{7.27}$$

with property $\pounds\left[c\right] = 0$, where c is a constant.

We now define a non-linear operator as

$$\begin{aligned} N\left[\phi\left(n,t;q\right)\right] &= L\left[\phi_n\left(n,t;q\right)\right] - \frac{1}{s}\left(-c\coth\left(d\right) + c\tanh\left(dn\right)\right) \\ &\quad - \frac{1}{s^{\alpha}}L\left[\phi_n\left(\Phi_n - \Phi_{n-1}\right)\right]. \\ N\left[\phi\left(n,t;q\right)\right] &= L\left[\Phi_n\left(n,t;q\right)\right] - \frac{1}{s}\left(-c\coth\left(d\right) - c\tanh\left(dn\right)\right) \\ &\quad - \frac{1}{s^{\alpha}}L\left[\Phi_n\left(\phi_{n+1} - \phi_n\right)\right]. \end{aligned} \tag{7.28}$$

Using the earlier definition, with assumption $H(n,t) = 1$, we construct the zeroth-order deformation equation

$$(1-q)\,\pounds\left[\phi_j\left(n,\mathrm{t};q\right) - u_{n,0}(n,t)\right] = qhN\left[\phi_j\left(n,t;q\right)\right], \quad j = 1,2. \tag{7.29}$$

Obviously, when $q = 0$ and $q = 1$,

$$\phi_j\left(n,t;0\right) = u_{n,0}(n,\mathrm{t}),\ \phi_j\left(n,t;1\right) = u_n(n,\mathrm{t}), \quad j = 1,2. \tag{7.30}$$

Thus, we obtain the mth-order deformation equations, respectively, as

$$\begin{aligned} L\left[u_{n,m}\left(n,\mathrm{t}\right) - \chi_m u_{n,m-1}(n,\mathrm{t})\right] &= hR_m(\vec{u}_{n,m-1},n,t), \\ L\left[v_{n,m}\left(n,\mathrm{t}\right) - \chi_m v_{n,m-1}(n,\mathrm{t})\right] &= hR_m(\vec{u}_{n,m-1},n,t), \end{aligned} \tag{7.31}$$

where

$$\begin{aligned} R_m\left(\vec{u}_{n,m-1},n,t\right) &= L[u_{n,m-1}] - \frac{(1-\chi_m)}{s}\left(-c\coth\left(d\right) + c\tanh\left(dn\right)\right) \\ &\quad - s^{-\alpha}L\left[\sum_{k=0}^{m-1} u_{n,\,m-k-1}\left(v_{n,k} - v_{n-1,k}\right)\right], \quad m \geq 1 \\ R_m\left(\vec{v}_{n,m-1},n,t\right) &= L[v_{n,m-1}] - \frac{(1-\chi_m)}{s}\left(-c\coth\left(d\right) - c\tanh\left(dn\right)\right) \\ &\quad - s^{-\alpha}L\left[\sum_{k=0}^{m-1} v_{n,m-1-k}\left(u_{n+1,k} - u_{n,k}\right)\right], \quad m \geq 1 \end{aligned} \tag{7.32}$$

Now, the solutions of mth-order deformation Eq. (7.31) are, respectively,

$$\begin{aligned} u_{n,m}(n,\mathrm{t}) &= (\chi_m + h)u_{n,m-1} - \left(-c\coth\left(d\right) + c\tanh\left(dn\right)\right)h(1-\chi_m) \\ &\quad - hL^{-1}\left(s^{-\alpha}L\left[\sum_{k=0}^{m-1} u_{n,m-1-k}\left(v_{n,k} - v_{n-1,k}\right)\right]\right), \\ v_{n,m}(n,\mathrm{t}) &= (\chi_m + h)v_{n,m-1} - \left(-c\coth\left(d\right) - c\tanh\left(dn\right)\right)h(1-\chi_m) \\ &\quad - hL^{-1}\left(s^{-\alpha}L\left[\sum_{k=0}^{m-1} v_{n,m-1-k}\left(u_{n+1,k} - u_{n,k}\right)\right]\right), \end{aligned} \tag{7.33}$$

Using the initial condition $u_0(n,t) = -c\coth(d) + c\tanh(dn)$ and $v_0(n,t) = -c\coth(d) - c\tanh(dn)$ and the iterative scheme (7.33), we obtained various iterates

$$u_{n,1} = \frac{c^2 h t^\alpha}{\Gamma(1+\alpha)} \big(\coth d \tanh d(-1+n) - \coth d \tanh dn - \tanh dn \tanh d(-1+n) + \tanh^2 dn\big),$$

$$v_{n,1} = \frac{c^2 h t^\alpha}{\Gamma(1+\alpha)} \big(\coth d \tanh d(1+n) - \coth d \tanh dn + \tanh dn \tanh d(1+n) - \tanh^2 dn\big),$$

$$\begin{aligned} u_{n,2} = {} & \frac{c^2 h(1+h) t^\alpha}{\Gamma(1+\alpha)} \big(\coth d \tanh d(-1+n) - \coth d \tanh dn \\ & - \tanh dn \tanh d(-1+n) + \tanh^2 dn\big) + \frac{c^3 h^2 t^{2\alpha}}{\Gamma(1+2\alpha)} \\ & \times \big(\coth^2 d \tanh d(-1+n) - 2\coth^2 d \tanh dn \\ & - \tanh^2 dn \tanh d(-1+n) + 2\tanh^3 dn + \coth^2 d \tanh d(1+n) \\ & - \tanh^2 dn \tanh d(1+n)\big), \end{aligned}$$

$$\begin{aligned} v_{n,2} = {} & \frac{c^2 h(1+h) t^\alpha}{\Gamma(1+\alpha)} \big(\coth d \tanh d(1+n) - \coth d \tanh dn \\ & + \tanh dn \tanh d(1+n) - \tanh^2 dn\big) + \frac{c^3 h^2 t^{2\alpha}}{\Gamma(1+2\alpha)} \\ & \times \big(-\coth^2 d \tanh d(-1+n) + 2\coth^2 d \tanh dn \\ & + \tanh^2 dn \tanh d(-1+n) - 2\tanh^3 dn - \coth^2 d \tanh d(1+n) \\ & + \tanh^2 dn \tanh d(1+n)\big). \end{aligned}$$

Proceeding in this manner, the rest of the components of $u_{n,m}(n,t)$ and $v_{n,m}(n,t)$, $m \geq 3$ can be completely obtained, and the series solution are thus entirely determined. Finally, the solution of Eq. (7.23) is given as

$$u_n(n,t) = \sum_{m=0}^{\infty} u_{n,m}(n,t), \; v_n(n,t) = \sum_{m=0}^{\infty} v_{n,m}(n,t). \tag{7.34}$$

By taking $\hbar = -1$ and $\alpha = 1$ in the above expressions, we get exactly the same as those given by HAM by Liao [11]. However, mostly, the results given by the Adomian decomposition method and homotopy perturbation method converge to the corresponding numerical solutions in a rather small region. But, different from those two methods, the fractional HATM provides us with a simple way to adjust and control the convergence region of solution series by choosing a proper value for the auxiliary parameter $\hbar$.

Figures 7.1 and 7.2 show the graphical comparison between the exact and approximate solutions obtained by the proposed HATM with Adomian's polynomials. It is seen from Figures 7.1 and 7.2 that the solution obtained by the proposed method nearly identical to the known exact solutions. It is also seen that the approximate analytical solution obtained by a proposed method increases rapidly for $u(n,t)$ and decreases for $v(n,t)$ with the increases in t. Figures 7.3 and 7.4 show the approximate solutions for different values of $\alpha = 0.7$, $\alpha = 0.8$, $\alpha = 0.9$ and standard Toda lattice equations, i.e. $\alpha = 1$.

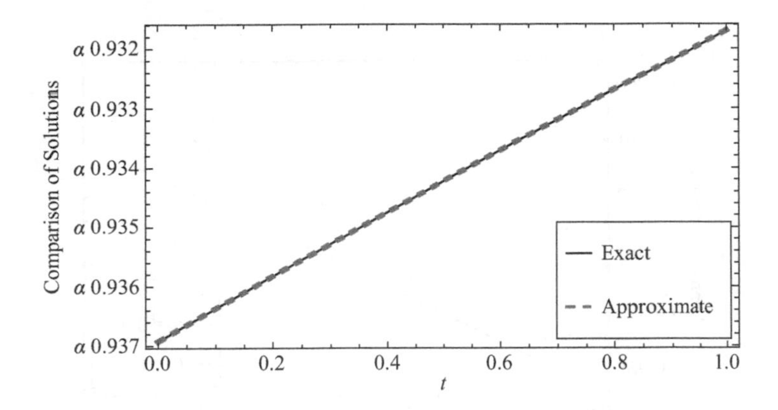

FIGURE 7.1
Plot of comparison between the exact $u(n,t)$ and approximate solutions $u_7(n,t)$ at $n = 120, c = d = 0.1$ and $\alpha = 1$.

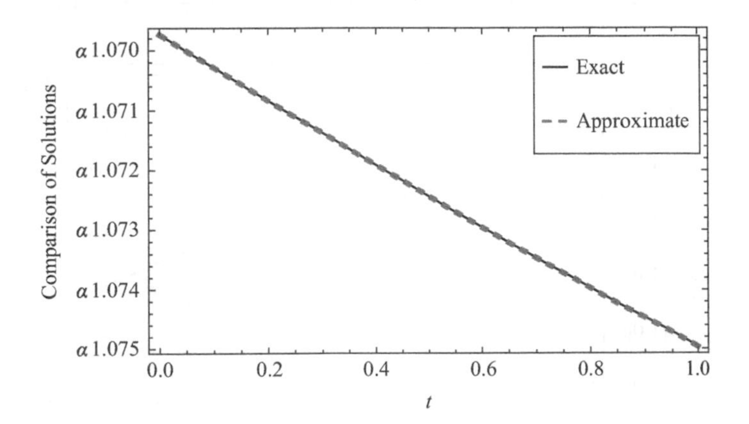

FIGURE 7.2
Plot of comparison between the exact $v(n,t)$ and approximate solutions $v_7(n,t)$ at $n = 120$, $c = d = 0.1$ and $\alpha = 1$.

The simplicity and accuracy of the proposed method are illustrated by computing the absolute error $E_7(u) = |u(n,t) - u_7(n,t)|$, $E_7(v) = |v(n,t) - v_7(n,t)|$, where $u(n,t)$, $v(n,t)$ are the exacts and $u_7(n,t)$, $v_7(n,t)$ is the seventh-order approximate solution of the fractional Toda lattice Eq. (7.23) obtained by truncating the respective solution series (7.34) at level $m = 7$. Figures 7.5 and 7.6 represent the absolute error curve, and it is seen from the figures that our approximate solutions obtained by fractional HATM converge very rapidly to the exact solutions in only seventh-order approximations, i.e. approximate solutions are very near to the exact solutions. It achieves a high level of accuracy. Of course, the accuracy level can be improved by introducing more terms of approximate solutions.

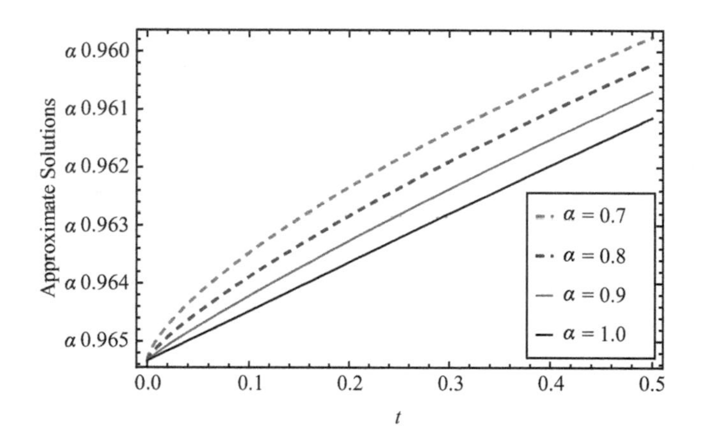

FIGURE 7.3
Plot of $u_7(n,t)$ vs. time t for different values of α at $n = 120$ and $c = d = 0.1$.

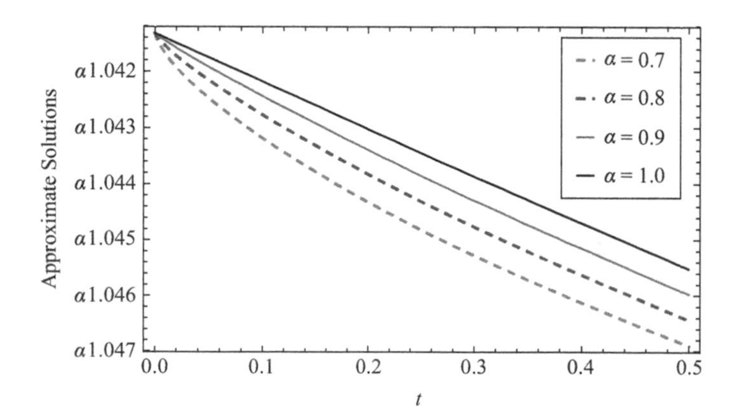

FIGURE 7.4
Plot of $v_7(n,t)$ vs. time t for different values of α at $n = 120$ and $c = d = 0.1$.

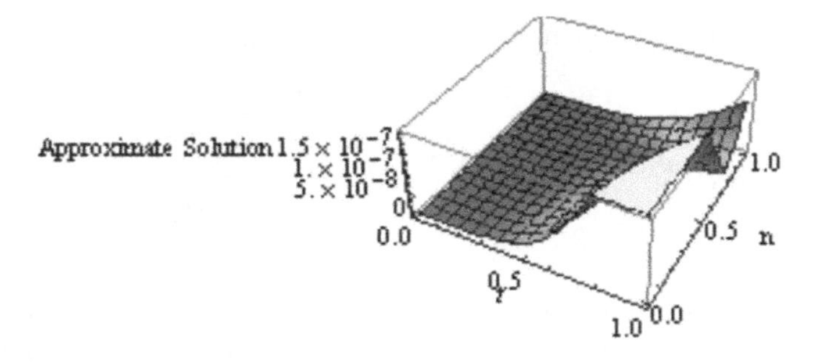

FIGURE 7.5
Plot of absolute error between the exact $u(n,t)$ and approximate solutions $u_7(n,t)$ at $c = d = 0.1$ and $\alpha = 1$.

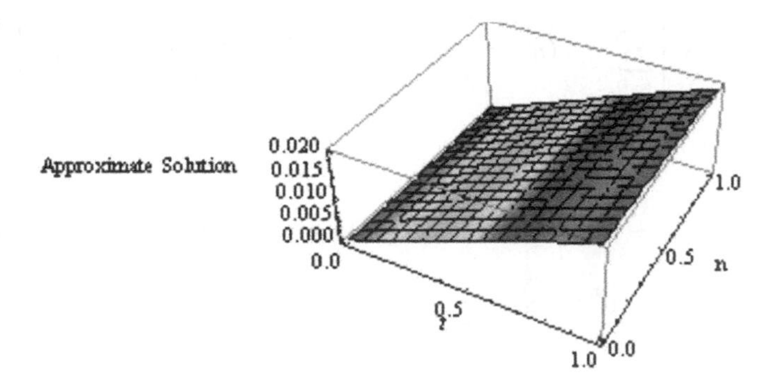

FIGURE 7.6
Plot of absolute error between the exact $v(n,t)$ and approximate solutions $v_7(n,t)$ at $c = d = 0.1$ and $\alpha = 1$.

There exist some methods to accelerate the convergence of series solution (7.34) and increase the convergence region. Among those, the homotopy-Padé technique is widely applied. The homotopy-Padé technique is combination of the Padé technique and the HAM [18]. The corresponding $[m,n]$ Padé approximant for $\phi(n,t;q)$ about the embedding parameter q is expressed by

$$\phi(n,t;q) = u_0(n,t) + \sum_{m=1}^{\infty} u_m(n,t)q^m = \frac{\sum\limits_{k=0}^{m} A_{m,k}(n,t)q^k}{\sum\limits_{k=0}^{n} B_{m,k}(n,t)q^k}, \tag{7.35}$$

where $A_{m,k}(n,t)$ and $B_{m,k}(n,t)$ are polynomial and entirely determined by the set of approximations

$$u_0(x,t),\ u_1(x,t),\ u_2(x,t), \ldots, u_{m+n}(x,t).$$

Putting $q = 1$, the $[m, n]$ homotopy Padé approximant is obtained as

$$\phi(n, t; 1) = u_0(n, t) + \sum_{m=1}^{\infty} u_m(n, t) = \frac{\sum_{k=0}^{m} A_{m,k}(n, t)}{\sum_{k=0}^{n} B_{m,k}(n, t)}. \tag{7.36}$$

The homotopy-Padé approximation enlarges the convergence region and rate of solution series given by HAM. This is valid even when the time span is increased.

At the mth order of approximation, one can define the exact square residual error as

$$\Delta_m = \int_0^1 \left(N \left[\sum_{i=0}^{m} u_{n,i}(n, t) \right] \right)^2 \mathrm{d}t, \ \Delta_m = \int_0^1 \left(N \left[\sum_{i=0}^{m} v_{n,i}(n, t) \right] \right)^2 \mathrm{d}t. \tag{7.37}$$

where

$$N\left[u_n(n, t)\right] = \frac{\mathrm{d}^\alpha u_n(n, t)}{\mathrm{d}t^\alpha} - u_n(n, t)v_n(n, t) + u_n(n, t)v_{n-1}(n, t),$$

$$N\left[v_n(n, t)\right] = \frac{\mathrm{d}^\alpha v_n(n, t)}{\mathrm{d}t^\alpha} - v_n(n, t)u_{n+1}(n, t) + u_n(n, t)v_n(n, t).$$

However, the exact square residual error Δ_m defined earlier needs too much central processing unit (CPU) time to calculate, even if the order of approximation is not very high.

Thus, to overcome this difficulty, i.e. to decrease the CPU time, we use here the so-called averaged residual error defined by

$$E_m = \frac{1}{5} \sum_{j=0}^{5} \left(N \left[\sum_{i=0}^{m} u_{n,i}\left(n, \frac{3j}{5}\right) \right] \right)^2,$$

$$E_m = \frac{1}{5} \sum_{j=0}^{5} \left(N \left[\sum_{i=0}^{m} v_{n,i}\left(n, \frac{3j}{5}\right) \right] \right)^2.$$

$$E_m = \frac{1}{5} \sum_{j=0}^{5} \left(N \left[\sum_{i=0}^{m} v_{n,i}\left(n, \frac{3j}{5}\right) \right] \right)^2. \tag{7.38}$$

The optimal value of h can be obtained by means of minimizing the so-called averaged residual error E_m defined by earlier, corresponding to the non-linear algebraic equation $\frac{\mathrm{d}E_m}{\mathrm{d}h} = 0$.

7.4 Numerical Result and Discussion

As pointed out by Liao, the convergence and rate of approximation for the HAM solution strongly depend on the value of auxiliary parameter $\hbar$ even if the initial approximation $u_0(x,t)$, the auxiliary linear operator L and the auxiliary function $H(x,t)$ are given, and we still have great freedom to choose the value of the auxiliary parameter $\hbar$. So, we can mathematically describe the so-called $\hbar$-curve that it provides us with an additional way to conveniently adjust and control the convergence region and rate of solution series and the valid regions of $\hbar$ to gain a convergent solution series. When the valid region of $\hbar$ is a horizontal line segment, then the solution is converged. Without loss of generality, we consider here the two cases of $n = 80$ and $n = 120$. Note that the convergence region and rate of the solution series (7.34) are determined by the auxiliary parameter $\hbar$. We can always find, by means of plotting the so-called $\hbar$-curves, a valid region of $\hbar$ to ensure that the solution series (7.34) converges.

Figures 7.7 and 7.8 show the $\hbar$-curve obtained from the seventh-order HATM approximation solution. In our study, it is obvious from Figures 7.7 and 7.8 that the acceptable range of auxiliary parameter $\hbar$ is $-2.00 \leq \hbar < 0$. We still have the freedom to choose the auxiliary parameter according to the $\hbar$ curve.

From Tables 7.1 to 7.4, it is observed that the values of the approximate solution at different grid points obtained by the new proposed method are very much close to the exact solution, with high accuracy at level $m = 7$. It can also be noted that the accuracy increases as the value of n increases.

The optimal value of h and the value of averaged residual error E_m are given in Tables 7.5–7.16 for the different order of approximation in the case of

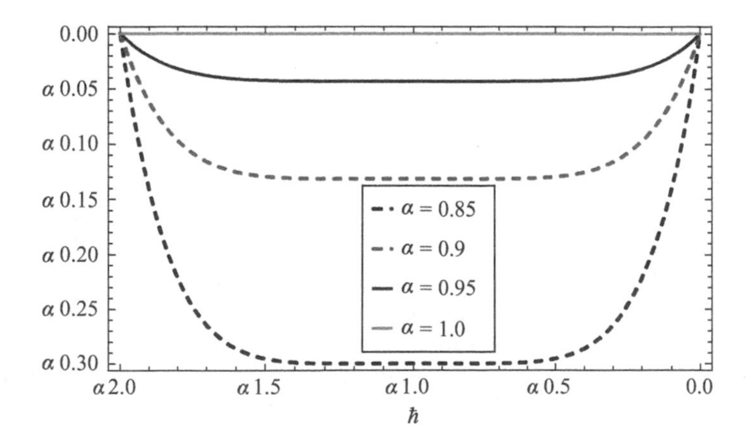

FIGURE 7.7
Plot of $\hbar$ curves for different values of α at $t = 2$.

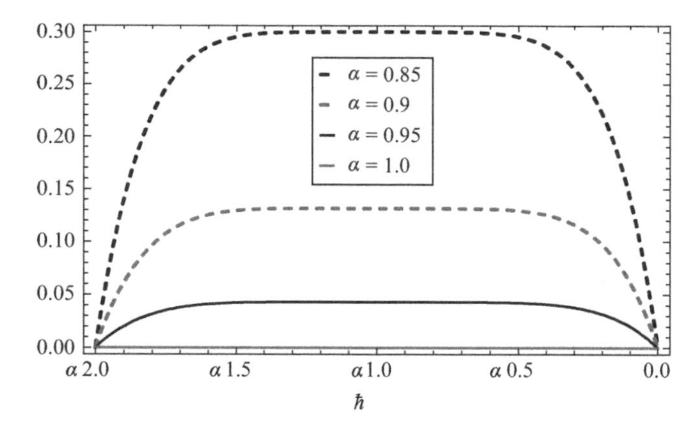

FIGURE 7.8
Plot of $\hbar$ curves for different values of α at $t = 2$.

both $u(n,t)$ and $v(n,t)$. It is clear from Tables 7.5, 7.7 and 7.9 that the optimal values of h are -0.71380, -0.83639 and -0.69988 for order of derivatives $\alpha = 1, 0.875$ and 0.75, respectively, for $n = 80$ in case of $u(n,t)$. Again, in Tables 7.6, 7.8 and 7.10, the optimal values of h are -0.71321, -0.70027 and -0.70143 for the order of derivatives $\alpha = 1, 0.875$ and 0.75, respectively, for $n = 120$ in the case of $u(n,t)$. Similarly, we have the optimal values in Tables 7.11, 7.13, 7.15 and 7.12, 7.14, 7.16 for $v(n,t)$. It is also observed from the table that as the value of n increases, the value of E_m gives more accuracy, and the optimal value of h deviates from $h = -1$ for both $u(n,t)$ and $v(n,t)$. Also, to measure the accuracy of the obtained results, mean square error (MSE) is computed for time-fractional Toda lattice equation, where MSE $= \frac{1}{n}\sum_{i=1}^{n}(u_i - u_{7,i})^2$.

It is clear from Tables 7.17 and 7.18 that our method converges rapidly towards the exact one.

7.5 Concluding Remarks

This chapter develops an effective and new amagalation of HAM and Laplace transform with Adomian's polynomial for fractional Toda lattice equations. The fractional derivatives here are described in the Caputo sense. This method gives more realistic series solutions that converge very rapidly in physical problems. We have discussed the methodology for the construction of these schemes and studied their performance on a problem. An excellent agreement is achieved. The solution is rapidly convergent by utilizing HAM by modification of Laplace operator and Adomian's polynomials. It may be concluded that the HATM methodology is very powerful and efficient in finding approximate solutions as well as analytical solutions of many fractional physical models arising in science, engineering and other physical phenomena.

TABLE 7.1
Comparison between the Exact and Seventh-Term Approximation Solutions

t	**Exact Solution**	**Appr. Solution** $\alpha = 1$	**Appr. Solution** $\alpha = 0.9$	$E_7(u) = \vert u - u_7 \vert$ **for** $\alpha = 1$	$E_7(u) = \vert u - u_7 \vert$ **for** $\alpha = 0.9$
5	−0.9033311353	−0.9033311353	−0.9033311351	-2.183×10^{-22}	-1.35127×10^{-22}
10	−0.9033311349	−0.9033311349	−0.9033311346	-2.42473×10^{-25}	-1.8792×10^{-22}
15	−0.9033311344	−0.9033311342	−0.9033311342	-8.22564×10^{-25}	-2.1316×10^{-22}
20	−0.9033311340	−0.9033311340	−0.9033311337	$-1.955866 \times 10^{-25}$	-2.2160×10^{-22}
25	−0.9033311336	−0.9033311336	−0.9033311333	-3.87334×10^{-25}	-2.183×10^{-22}

Comparison between exact and seventh-term approximation solutions for $u_n(n,t)$. Taking $c = d = 0.1$, $n = 80$, $h = -1$, $\alpha = 1$ and 0.9.

TABLE 7.2
Comparison between the Exact and Seventh-Term Approximation Solutions

t	**Exact Solution**	**Appr. Solution** $\boldsymbol{\alpha = 1}$	**Appr. Solution** $\boldsymbol{\alpha = 0.9}$	$\boldsymbol{E_7(u) = \lvert u - u_7 \rvert}$ **for** $\boldsymbol{\alpha = 1}$	$\boldsymbol{E_7(u) = \lvert u - u_7 \rvert}$ **for** $\boldsymbol{\alpha = 0.9}$
5	−1.1033310912	−1.1033310912	−1.10333109129	3.01981×10^{-25}	1.35214×10^{-22}
10	−1.1033310916	−1.1033310916	−1.10333109178	2.42251×10^{-25}	1.88495×10^{-22}
15	−1.1033310920	−1.1033310920	−1.10333109224	8.20677×10^{-25}	2.14899×10^{-22}
20	−1.1033310925	−1.1033310924	−1.1033310926	1.95266×10^{-25}	2.25426×10^{-22}
25	−1.1033310929	−1.1033310929	−1.10333109308	3.8296×10^{-25}	2.25361×10^{-22}

Comparison between the exact and seventh-term approximation solutions for $v_n(n,t)$. Taking $c = d = 0.1$, $n = 80$, $h = -1$, $\alpha = 1$ and 0.9.

TABLE 7.3
Comparison between the Exact and Seventh-Term Approximation Solutions

t	**Exact Solution**	**Appr. Solution** $\alpha = 1$	**Appr. Solution** $\alpha = 0.9$	$E_7(u) = \lvert u - u_7 \rvert$ **for** $\alpha = 1$	$E_7(u) = \lvert u - u_7 \rvert$ **for** $\alpha = 0.9$
5	−0.9033311132	−0.9033311132	−0.9033311132	1.44329×10^{-27}	-4.27936×10^{-26}
10	−0.9033311132	−0.9033311132	−0.9033311132	5.88418×10^{-27}	-5.31797×10^{-26}
15	−0.9033311132	−0.9033311132	−0.9033311132	1.35447×10^{-27}	-5.06262×10^{-26}
20	−0.9033311132	−0.9033311132	−0.9033311132	2.42029×10^{-27}	-3.94129×10^{-26}
25	−0.9033311132	−0.9033311132	−0.9033311132	3.83027×10^{-27}	-2.12053×10^{-26}

Comparison between the exact and seventh-term approximation solutions for $u_n(n,t)$. Taking $c = d = 0.1$, $n = 120$, $h = -1$, $\alpha = 1$ and 0.9.

TABLE 7.4
Comparison between the Exact and Seventh-Term Approximation Solutions

t	**Exact Solution**	**Appr. Solution** $\alpha = 1$	**Appr. Solution** $\alpha = 0.9$	$E_7(v) = \lvert v - v_7 \rvert$ **for** $\alpha = 1$	$E_7(v) = \lvert v - v_7 \rvert$ **for** $\alpha = 0.9$
5	−1.1033311132	−1.1033311132	−1.1033311132	2.22045×10^{-30}	4.50751×10^{-28}
10	−1.1033311132	−1.1033311132	−1.1033311132	2.22045×10^{-30}	6.32827×10^{-28}
15	−1.1033311132	−1.1033311132	−1.1033311132	0	7.23865×10^{-28}
20	−1.1033311132	−1.1033311132	−1.1033311132	8.88178×10^{-30}	7.52731×10^{-28}
25	−1.1033311132	−1.1033311132	−1.1033311132	8.88178×10^{-30}	7.57172×10^{-28}

Comparison between the exact and seventh-term approximation solutions for $v_n(n,t)$. Taking $c = d = 0.1$, $n = 120$, $h = -1$, $\alpha = 1$ and 0.9.

TABLE 7.5
Optimal Value of h for $c = d = 0.1,\ n = 80$ When $\alpha = 1$ for $u(n,t)$

Order of Approximation	Optimal Value of h	Value of E_m
1	−1.01010	3.89047×10^{-19}
2	−0.94253	2.55609×10^{-17}
3	−0.71380	1.48295×10^{-18}

TABLE 7.6
Optimal Value of h for $c = d = 0.1,\ n = 120$ When $\alpha = 1$ for $u(n,t)$

Order of Approximation	Optimal Value of h	Value of E_m
1	−1.01000	6.93042×10^{-26}
2	−0.78137	2.56611×10^{-25}
3	−0.71321	1.67118×10^{-25}

TABLE 7.7
Optimal Value of h for $c = d = 0.1,\ n = 80$ When $\alpha = 0.875$ for $u(n,t)$

Order of Approximation	Optimal Value of h	Value of E_m
1	−1.13363	5.96133×10^{-19}
2	−0.76264	3.41132×10^{-18}
3	−0.83639	2.28271×10^{-18}

TABLE 7.8
Optimal Value of h for $c = d = 0.1,\ n = 120$ When $\alpha = 0.875$ for $u(n,t)$

Order of Approximation	Optimal Value of h	Value of E_m
1	−1.13363	6.70597×10^{-26}
2	−0.76266	3.83776×10^{-25}
3	−0.70027	2.56538×10^{-25}

TABLE 7.9
Optimal Value of h for $c = d = 0.1,\ n = 80$ When $\alpha = 0.75$ for $u\,(n, t)$

Order of Approximation	Optimal Value of h	Value of E_m
1	−1.28602	1.12123×10^{-18}
2	−0.75226	4.73546×10^{-18}
3	−0.69988	3.34029×10^{-18}

TABLE 7.10
Optimal Value of h for $c = d = 0.1,\ n = 120$ When $\alpha = 0.75$ for $u\,(n, t)$

Order of Approximation	Optimal Value of h	Value of E_m
1	−1.28586	1.26072×10^{-25}
2	−0.75225	5.3273×10^{-25}
3	−0.70143	3.7495×10^{-25}

TABLE 7.11
Optimal Value of h for $c = d = 0.1,\ n = 80$ When $\alpha = 1$ for $v\,(n, t)$

Order of Approximation	Optimal Value of h	Value of E_m
1	−1.01010	6.15985×10^{-19}
2	−0.781387	2.28057×10^{-18}
3	−1.19047	1.22019×10^{-18}

TABLE 7.12
Optimal Value of h for $c = d = 0.1,\ n = 120$ When $\alpha = 1$ for $v\,(n, t)$

Order of Approximation	Optimal Value of h	Value of E_m
1	−1.01015	6.93227×10^{-26}
2	−0.78139	2.56681×10^{-25}
3	−1.19128	1.37519×10^{-25}

TABLE 7.13
Optimal Value of h for $c = d = 0.1,\ n = 80$ When $\alpha = 0.875$ for $v(n,t)$

Order of Approximation	Optimal Value of h	Value of E_m
1	−1.13363	5.96133×10^{-19}
2	−0.76264	3.41132×10^{-18}
3	−1.27309	1.8922×10^{-18}

TABLE 7.14
Optimal Value of h for $c = d = 0.1,\ n = 120$ When $\alpha = 0.875$ for $v(n,t)$

Order of Approximation	Optimal Value of h	Value of E_m
1	−1.13369	6.71118×10^{-26}
2	−0.76267	3.83906×10^{-25}
3	−1.2725	2.12656×10^{-25}

TABLE 7.15
Optimal Value of h for $c = d = 0.1,\ n = 80$ When $\alpha = 0.75$ for $v(n,t)$

Order of Approximation	Optimal Value of h	Value of E_m
1	−1.28602	1.12123×10^{-18}
2	−0.75226	4.73546×10^{-18}
3	−1.35093	2.8707×10^{-18}

TABLE 7.16
Optimal Value of h for $c = d = 0.1,\ n = 120$ When $\alpha = 0.75$ for $v(n,t)$

Order of Approximation	Optimal Value of h	Value of E_m
1	−1.28602	1.26191×10^{-25}
2	−0.75230	5.32819×10^{-25}
3	−1.35051	3.22791×10^{-25}

TABLE 7.17
MSE for $u_n(n,t)$ When $h=-1,\ c=d=0.1$ and $t=5$

n	MSE for $\alpha=1$	MSE for $\alpha=0.9$
5	1.65434×10^{-10}	6.90876×10^{-9}
10	2.56214×10^{-12}	3.89704×10^{-11}
15	5.78543×10^{-14}	4.56721×10^{-13}
20	9.06732×10^{-16}	8.74513×10^{-15}

TABLE 7.18
MSE for $v_n(n,t)$ When $h=-1,\ c=d=0.1$ and $t=5$

n	MSE for $\alpha=1$	MSE for $\alpha=0.9$
5	2.87013×10^{-8}	7.00123×10^{-7}
10	7.91234×10^{-10}	6.20201×10^{-9}
15	9.97341×10^{-12}	4.87610×10^{-11}
20	3.02021×10^{-14}	0.01276×10^{-13}

Acknowledgements

The second author Dr Sunil Kumar would like to acknowledge the financial support received from the National Board for Higher Mathematics, Department of Atomic Energy, Government of India (Approval No. 2/48(20)/2016/NBHM(R.P.)/R and D II/1014).

References

[1] West, B. J., Bolognab, M. and Grigolini, P., *Physics Operator*, Springer, 2003 (in New York).

[2] Oldham, K. B. and Spanier, J., *The Fractional Calculus*, Academic Press, 1974 (in New York).

[3] Miller, K. S. and Ross, B., *An Introduction to the Fractional Integrals and Derivatives—Theory and Application*, Wiley, 1993 (in New York).

[4] Samko, S. G., Kilbas, A. A. and Marichev, O. I., *Fractional Integrals and Derivatives Theory and Applications*, Gordon and Breach, 1993 (in New York).

[5] Hilfer, R., *Application of Fractional Calculusin Physics*, World Scientific, 2000 (in Germany).

[6] Podlubny, I., *Fractional Differential Equations*, Academic, 1999 (in New York).

[7] Zaslavsky, G. M., *Hamiltonian Chaos and Fractional Dynamics*, Oxford University Press, 2005 (in Oxford).

[8] Kilbas, A. A., Srivastava, H. M. and Trujillo, J. J., *Theory and Applications of Fractional Differential Equations*, Elsevier, 2006 (in Duivendrecht, The Netherlands).

[9] Magin, R. L., *Fractional Calculus in Bio-engineering*, Begell House Publisher, Inc., 2006 (in Danbury, CT).

[10] Gorenflo, R. and Mainardi, F., *Fractional Calculus: Integral and Differential Equations of Fractional Orders, Fractals and Fractional Calculus in Continuum Mechanics*, Springer Verlag, 1997 (in New York).

[11] Liao, S. J., The proposed homotopy analysis technique for the solution of nonlinear problems, PhD thesis, Shanghai Jiao Tong University, 1992.

[12] Liao, S. J., *Beyond Perturbation: Introduction to the Homotopy Analysis Method*, CRC Press, 2003 (Boca Raton, FL).

[13] Liao, S. J. (2004). On the homotopy analysis method for nonlinear problems. *Applied Mathematics and Computation*, 147, 499–513.

[14] Liao, S. J. (2009). Notes on the homotopy analysis method: Some definition and theorems. *Communication Nonlinear Science and Numerical Simulation*, 14, 983–997.

[15] Liao, S. J. (1997). Homotopy analysis method: A new analytical technique for nonlinear problems. *Communication Nonlinear Science and Numerical Simulation*, 2, 95–100.

[16] Liao, S. J. (2005). A new branch of solutions of boundary-layer flows over an impermeable stretched plate. *International Journal of Heat and Mass Transfer*, 48, 2529–2539.

[17] Liao, S. J. and Pop, I. (2004). Explicit analytic solution for similarity boundary layer equations. *International Journal of Heat and Mass Transfer*, 47, 75–78.

[18] Vishal, K., Kumar, S. and Das, S. (2012). Application of homotopy analysis method for fractional swift Hohenberg equation-revisited. *Applied Mathematical Modeling*, 36(8), 3630–3637.

[19] Zhang, X., Tang, B. and He, Y. (2011). Homotopy analysis method for higher-order fractional integro-differential equations. *Computer Mathematics and Applications*, 62, 3194–3203.

[20] Rashidi, M. M. and Pour, S. A. M. (2010). Analytic approximate solutions for unsteady boundary-layer flow and heat transfer due to a stretching sheet by homotopy analysis method. *Nonlinear Analysis Modeling and Control*, 15(1), 83–95.

[21] Rashidi, M. M., Domairry, G. and Dinarvand, S. (2009). The homotopy analysis method for explicit analytical solutions of Jaulent Miodek equations. *Numerical Methods for Partial Differential Equations*, 25(2), 430–439.

[22] Khan, N. A., Jamil, M. and Ara, A. (2012). Approximate solution of time fractional Schrodinger equation via homotopy analysis method. *ISRN Mathematical Physics*, Article ID 197068, 11 pages.

[23] Abbasbandy, S. (2008). Homotopy analysis method for generalized Benjamin-Bona-Mahony equation. *Zeitchrift fur Angewandte Mathematik and Physik*, 59, 51–62.

[24] Abbasbandy, S. and Zakaria, F. S. (2008). Soliton solutions for the fifth-order KdV equation with the homotopy analysis method. *Nonlinear Dynamics*, 51, 83–87.

[25] Abbasbandy, S. (2010). Homotopy analysis method for the Kawahara equation. *Nonlinear Analalysis: Real World Applications*, 11, 307–312.

[26] Kumar, S., Kumar, A. and Baleanu, D. (2016). Two analytical methods for time-fractional non-linear coupled Boussinesq-Burgers equations arise in propagation of shallow water waves. *Nonlinear Dynamic*, 85(2), 699–715.

[27] Kumar, S., Kumar, A. and Argyros, I. K. (2017). A new analysis for the Keller-Segel model of fractional order. *Numerical Algorithm*, 75(1), 213–228.

[28] Khader, M. M., Kumar, S. and Abbasbandy, S. (2013). New homotopy analysis transform method for solving the discontinued problems arising in nanotechnology. *Chinese Physics B*, 22(11), 110201.

[29] Arife, A. S., Vanani, S. K., and Soleymani, F. (2012). The Laplace homotopy analysis method for solving a general fractional diffusion equation arising in nano-hydrodynamics. *Journal of Computational and Theoretical Nanoscience*, 10, 1–4.

[30] Kumar, S. (2013). An analytical algorithm for nonlinear fractional Fornberg-Whitham equation arising in wave breaking based on a new iterative method. *Alexander Engineering Journal*, 52(4), 813–819.

[31] Kumar, S. (2013). A new analytical modelling for telegraph equation via Laplace transforms. *Applied Mathematical Modeling*, 38(13), 3154–3163.

[32] Mokhtari, R. (2002). Variational iteration method for solving nonlinear differential-difference equations. *International Journal of Nonlinear Science and Numerical Simulation*, 9(1), 19–24.

[33] Adomian, G. (1988). A review of the decomposition method in applied mathematics. *Journal of Mathematical Analysis and Application*, 135, 501–544.

[34] Cheng, J., Zhu, S. P. and Liao, S. J. (2010). An explicit series approximation to the optimal exercise boundary of American put options. *Communication Nonlinear Science and Numerical Simulation*, 15, 1148–1158.

[35] Liao, S. J. (2016). An optimal homotopy-analysis approach for strongly nonlinear differential equations. *Communication Nonlinear Science and Numerical Simulation*, 15, 2003–2016.

[36] Liao, S. J. (2012). *Homotopy Analysis Method in Nonlinear Differential Equations*, Higher Education Press, Springer, 2012 (in Beijing).

[37] Rashidi, M. M., Freidoonimehr, N., Hosseini, A., Bég, O. A. and Hung, T. K. (2014). Homotopy simulation of nanofluid dynamics from a non-linearly stretching isothermal permeable sheet with transpiration. *Meccanica*, 1–14, 49, 469–482.

[38] Vishal, K., Das, S., Ong, S. H. and Ghosh, P. (2013). On the solution of fractional Swift Hohenberg equation with dispersion. *Applied Mathematical and Computational*, 269, 5792–5801.

[39] Liao, S. J. (2010). An Optimal Homotopy-analysis approach for strongly nonlinear differential equation. *Communication Nonlinear Science and Numerical Simulation*, 15(8), 2003–2016.

[40] Biswas, A., Al-Amr, M. O., Rezazadeh, H., Mirzazadeh, M., Eslami, M., Zhou, Q., Moshokoa, S. P. and Belic, M. (2018). Resonant optical solitons with dual–power law nonlinearity and fractional temporal evolution. *Optik*, 165, 233–239.

[41] Qasim, A. F. and Al-Amr, M. O. (2018). Approximate solution of the Kersten-Krasil'shchik coupled Kdv-MKdV system via reduced differential transform method. *Eurasian Journal of Science & Engineering*, 4(2), 1–9.

[42] Al-Sawoor, A. J. and Al-Amr, M. O. (2014). A new modification of variational iteration method for solving reaction-diffusion system with fast reversible reaction. *Journal of Egyptian Mathematical Society*, 22(3), 396–401.

[43] Al-Amr, M. O. and El-Ganaini, S. (2017). New exact traveling wave solutions of the (4+1)-dimensional Fokas equation. *Computer Mathematics and Application*, 74, 1274–1287.

[44] Al-Sawoor, A. J. and Al-Amr, M. O. (2012). Numerical solution of a reaction-diffusion system with fast reversible reaction by using Adomian's decomposition method and He's variational iteration method. *Al-Rafidain Journal of Computational Science and Mathematics*, 9(2), 243–257.

[45] Al-Amr, M. O. (2015). Exact solutions of the generalized (2+1)-dimensional nonlinear evolution equations via the modified simple equation method. *Computer Mathematics and Application*, 69(5), 390–397.

[46] Al-Amr, M. O. (2014). New applications of reduced differential transform method. *Alexandria Engineering Journal*, 53(1), 243–247.

8

Fractional Model of a Hybrid Nanofluid

Muhammad Saqib and Sharidan Shafie
Universiti Teknologi Malaysia JB

Ilyas Khan
Majmaah University

CONTENTS

8.1 Introduction

In recent decades, it is acknowledged that fractional operators are appropriate tools for differentiation as compared to the local differentiation, particularly in physical real-world problems. These fractional operators can be constructed by the convolutions of the local derivative with kernel of fractional operators. Various kernels of fractional operators have been suggested in the literature, but the most common is the power law kernel ($x^{-\alpha}$) that is used in the construction of Reimann–Liouville and Caputo fractional operators [1]. Caputo and Fabrizio used exponential decay law $\exp(-\alpha x)$ in the construction of Caputo–Fabrizio fractional operator [2]. Atangana and Baleanu developed fractional operators in Caputo and Reimann–Liouville sense using the generalized Mittag-Leffler law $E_\alpha(-\phi x^\alpha)$ as a kernel [3]. All these fractional operators have some shortcomings and challenges, but at the same time, this area is growing fast, and researchers devoted their attention to this field, see [4–7] and the references therein.

It is important to mention here that fractional order calculus has many applications in almost every field of science and technology, which includes diffusion, relaxation process, control, electrochemistry and viscoelasticity [8,9]. Zafar and Fetecau [10] applied Caputo–Fabrizio fractional derivative to the flow of Newtonian viscous fluid flowing over an infinite vertical plate. Martis et al. [11] analyzed the flow of fractional Maxwell's fluid. According to their report, the fractional results showed an excellent agreement with experimental work by adjusting the fractional parameter. Alkahtani and Atangana [12] used different fractional operators to analyze the memory effect in a potential energy field caused by a charge. They presented some novel numerical approaches to the solutions of a fractional system of equations. Veiru et al. [13] presented exact solutions for the time fraction model of viscous fluid flow near a vertical plate, taking into consideration mass diffusion and Newtonian heating. Abro et al. [14] presented exact analytical solutions for the flow of Oldroyd-B fluid in a horizontal circular pipe. Jain [15] introduced a novel and powerful numerical scheme and implemented to different fractional order differential equations. Some other interesting and significant studies on fractional derivatives can be found in [16–20] and the references therein.

Nanofluid is the innovation of nanotechnology to overcome the problems of heat transport in many engineering and industrial sectors. A detailed discussion on nanofluid with a list of applications is reported by Wang and Mujumdar [21] in the review paper. Sheikholeslami et al. [22] numerically studied the shape effect and external magnetic field effect on $F_3O_4-H_2O$ nanofluid inside a porous enclosure. Hussanan et al. [23] developed exact solutions for nanofluid with different nanoparticles for the unsteady flow of micropolar fluid. The literature of nanofluid exponentially increased and reached to a next level by introducing hybrid nanofluids that are suspensions of two or more types of nanoparticles in the composite form with low concentration. Hybrid nanofluids are introduced to overcome the drawbacks of single nanoparticle suspensions and connect the synergetic effect of nanoparticles. Hybrid nanofluid is branded to further improve the thermal conductivity, heat transport ability which leads to industrial and engineering applications with low cost [24]. Hussain et al. [25] carried out entropy generation analysis on hybrid nanofluid in a cavity. Farooq [26] presented a numerical study on hybrid nanofluid, keeping into consideration suction/injection, entropy generation and viscous dissipation.

In the existing literature, experimental, theoretical and numerical studies on hybrid nanofluid are very limited. A study on a hybrid nanofluid fluid with exact solutions and Caputo-Fabrizio fractional derivative does not exist. So, there is an urgent need to contribute to the literature of hybrid nanofluid using the application of fractional differential equations. Motivated from the earlier discussion, this chapter focused on the heat transfer in hybrid nanofluid in two vertical parallel plates using fractional derivative approach. A water-based hybrid nanofluid is characterized here with composite hybrid nanoparticles of copper Cu and alumina Al_2O_3. The fractional Brinkman-type fluid model with physical initial and boundary conditions is considered for the flow phenomena. The Laplace transform technique is used to obtain exact analytical solutions for velocity and temperature profiles. Using the properties of Caputo–Fabrizio fractional derivative, the obtained solutions are reduced to classical form for $\alpha = 1$ and $\beta = 1$. To explore the physical aspect of the flow parameters, the solutions are numerically computed and plotted in different graphs with a physical explanation.

8.2 Problem's Description

Consider the unsteady free convection flow of generalized incompressible hybrid nanofluid in two infinite vertical parallel plates at a distance d. The plates are taken along the x-axis, and y-axis is chosen normal to it. At $t \leq 0$, the plates and fluid are at rest with ambient temperature T_0. After $t = 0^+$, the temperature of the plate at $y = d$ rises or lowers from T_0 to T_W' due to which free convection takes place. At this moment, the fluid starts motion in x-direction due to the temperature gradient that gives rise to buoyancy forces. The Brinkman-type fluid model is utilized to describe the flow phenomena of a hybrid nanofluid. Under the assumptions of Jan et al. [27], the governing equations of $Cu{-}Al_2O_3{-}Cu$ hybrid nanofluid are given by

$$\rho_{hnf}\left(\frac{\partial u\left(y,t\right)}{\partial t}+\beta_b^* u\left(y,t\right)\right)=\mu_{hnf}\frac{\partial^2 u\left(y,t\right)}{\partial y^2}+g\left(\rho\beta_T\right)_{hnf}\left(T\left(y,t\right)-T_0\right), \tag{8.1}$$

$$\left(\rho Cp\right)_{hnf}\frac{\partial T\left(y,t\right)}{\partial t}=k_{hnf}\frac{\partial^2 T\left(y,t\right)}{\partial y^2}+Q_0\left(T-T_0\right), \tag{8.2}$$

together with the following appropriate initial and boundary conditions

$$u\left(y,0\right)=0, T\left(y,0\right)=T_0, \forall y\geq 0, \tag{8.3}$$

$$\left.\begin{array}{ll} u\left(0,t\right)=0, T\left(0,t\right)=T_0, & \text{for}\, t>0, \\ u\left(d,t\right)=0, T\left(d,t\right)=T_W, & \text{for}\, t>0, \end{array}\right\}, \tag{8.4}$$

where ρ_{hnf} is the density, $u\left(y,t\right)$ is the velocity, β_b^* is Brinkmann parameter, μ_{hnf} is the dynamic viscosity, β_{hnf} is the volumetric thermal expansion, $\left(C_p\right)_{hnf}$ is the specific heat, $T\left(y,t\right)$ is the temperature, k_{hnf} is the thermal conductivity and Q_0 is the heat generation of the hybrid nanofluid. The numerical values of nanoparticles and base fluid are given in Table 8.1, whereas the modified form of mathematical expressions from the effective thermophysical properties is given in Table 8.2.

TABLE 8.1
Numerical Values of Thermophysical Properties of Base Fluid and Nanoparticles

Material	Base Fluid	Nanoparticles	
	H_2O	Al_2O_3	Cu
$\rho\,(\mathrm{kg/m^3})$	997.1	3970	8933
$C_p\,(\mathrm{J/kg\,K})$	4179	765	385
$K\,(\mathrm{W/m\,K})$	0.613	40	400
$\beta_T \times 10^{-5}(\mathrm{K^{-1}})$	21	0.85	1.67
Pr	6.2	–	–

8.3 Generalization of Local Model

In this section, the dimensional system is first transformed to a dimensionless form using non-similarity variables to reduce the number of variables and get rid of units. The dimensionless system is then artificially converted to time fractional form or generalized form using the Caputo–Fabrizio fractional operator [2]. It is worth to mention here that the fractional models are more general and convenient in the description of flow behaviour and memory effect. Moreover, the results obtained from the fractional model are additionally realistic, because by adjusting the fractional parameter, the obtained results can be compared with experimental data for an excellent agreement, as obtained by Markis et al. [11]. Now, introducing the following non-similarity dimensionless variables [28]

$$v = \frac{d}{\nu_f}u,\ \xi = \frac{y}{d},\ \tau = \frac{\nu_f}{d^2}t,\ \theta = \frac{T - T_0}{T_W - T_0},$$

into Eqs. (8.1)–(8.4) yield the following

$$a_0\left(\frac{\partial v(\xi,\tau)}{\partial \tau} + \beta_b v(\xi,\tau)\right) = a_1\frac{\partial^2 v(\xi,\tau)}{\partial \xi^2} + a_2 Gr\theta(\xi,\tau), \tag{8.5}$$

$$a_3 Pr\frac{\partial \theta(\xi,\tau)}{\partial \tau} = \lambda_{hnf}\frac{\partial^2 \theta(\xi,\tau)}{\partial \xi^2} + Q\theta(\xi,\tau), \tag{8.6}$$

$$v(\xi,0) = 0, \theta(\xi,0) = 0, \forall \xi \geq 0, \tag{8.7}$$

$$\left.\begin{array}{ll} v(0,\tau) = 0, \theta(0,\tau) = 0, & \text{for}\,\tau > 0, \\ v(1,\tau) = 0, \theta(1,\tau) = 1, & \text{for}\,t > 0, \end{array}\right\}, \tag{8.8}$$

where

$$\beta_b = \frac{d^2\beta_b^*}{\nu_f^2},\ Gr = \frac{d^3 g(\beta_T)_f}{\nu_f^2}(T_W - T_0),\ Pr = \frac{(\mu C_p)_f}{k_f},\ Q = \frac{d^2 Q_0}{k_f},$$

TABLE 8.2
Mathematical Expressions for Thermophysical Properties of Nanofluid and Hybrid Nanofluid

Nanofluid	Hybrid Nanofluid
$\rho_{nf} = (1-\phi)\,\rho_f + \phi\rho_s,$	$\rho_{hnf} = (1-\phi_{hnf})\,\rho_f + \phi_{\mathrm{Al_2O_3}}\rho_{\mathrm{Al_2O_3}} + \phi_{\mathrm{Cu}}\rho_{\mathrm{Cu}},$
$\mu_{nf} = \dfrac{\mu_f}{(1-\phi)^{2.5}},$	$\mu_{hnf} = \dfrac{\mu_f}{\{1-(\phi_{\mathrm{Al_2O_3}}+\phi_{\mathrm{Cu}})\}^{2.5}}$
$(\beta_T\rho)_{nf} = (1-\phi)\,(\beta_T\rho)_f + \phi\,(\beta_T\rho)_s,$	$(\rho\beta_T)_{hnf} = (1-\phi_{hnf})\,(\rho\beta_T)_f + \phi_{\mathrm{Al_2O_3}}\,(\rho\beta_t)_{\mathrm{Al_2O_3}} + \phi_{\mathrm{Cu}}\,(\rho\beta_T)_{\mathrm{Cu}}$
$(\rho C_p)_{nf} = (1-\phi)\,(\rho C_p)_f + \phi\,(\rho C_p)_s,$	$(\rho Cp)_{hnf} = (1-\phi_{hnf})\,(\rho Cp)_f + \phi_{\mathrm{Al_2O_3}}\,(\rho Cp)_{\mathrm{Al_2O_3}} + \phi_{\mathrm{Cu}}\,(\rho Cp)_{\mathrm{Cu}}$
$\dfrac{K_{nf}}{K_f} = \dfrac{k_s + 2k_f - 2\phi\,(k_s - k_f)}{k_s + 2k_f + \phi\,(k_s - k_f)},$	$\dfrac{k_{hnf}}{k_f} = \dfrac{\dfrac{\phi_{\mathrm{Al_2O_3}}k_{\mathrm{Al_2O_3}} + \phi_{\mathrm{Cu}}k_{\mathrm{Cu}}}{\phi_{hnf}} + 2k_f + 2\,(\phi_{\mathrm{Al_2O_3}}k_{\mathrm{Al_2O_3}} + \phi_{\mathrm{Cu}}k_{\mathrm{Cu}}) - 2k_f\phi_{hnf}}{\dfrac{\phi_{\mathrm{Al_2O_3}}k_{\mathrm{Al_2O_3}} + \phi_{\mathrm{Cu}}k_{\mathrm{Cu}}}{\phi_{hnf}} + 2k_f + (\phi_{\mathrm{Al_2O_3}}k_{\mathrm{Al2O3}} + \phi_{\mathrm{Cu}}k_{\mathrm{Cu}}) - k_f\phi_{hnf}}$

$$\lambda_{hnf} = \frac{k_{hnf}}{k_f}, \; a_0 = (1 - \phi_{hnf}) + \frac{\phi_{\mathrm{Al_2O_3}}\rho_{\mathrm{Al_2O_3}} + \phi_{\mathrm{Cu}}\rho_{\mathrm{Cu}}}{\rho_f},$$

$$a_1 = \frac{1}{\{1 - (\phi_{\mathrm{Al_2O_3}} + \phi_{\mathrm{Cu}})\}^{2.5}}, \; \phi_{hnf} = \phi_{\mathrm{Al_2O_3}} + \phi_{\mathrm{Cu}}$$

$$a_2 = (1 - \phi) + \frac{\phi_{\mathrm{Al_2O_3}} (\rho\beta_t)_{\mathrm{Al_2O_3}} + \phi_{\mathrm{Cu}} (\rho\beta_T)_{\mathrm{Cu}}}{(\rho\beta_T)_f},$$

$$a_3 = (1 - \phi) + \frac{\phi_{\mathrm{Al_2O_3}} (\rho Cp)_{\mathrm{Al_2O_3}} + \phi_{\mathrm{Cu}} (\rho Cp)_{\mathrm{Cu}}}{(\rho Cp)_f},$$

are the dimensionless Brinkman-type fluid parameter, the thermal Grashof number, the Prandtl number and heat generation parameter, respectively, where λ_{hnf}, a_0, a_1, a_2 and a_3 are the constant terms produced during the calculation. The time fractional form of Eqs. (8.5) and (8.6) in terms of Caputo–Fabrizio fractional derivatives is given by

$$\frac{a_0}{a_1} {}^{CF}D_\tau^\alpha v(\xi, \tau) + \frac{a_0}{a_1}\beta_b v(\xi, \tau) = \frac{\partial^2 v(\xi, \tau)}{\partial \xi^2} + \frac{a_2}{a_1} Gr\theta(\xi, \tau), \tag{8.9}$$

$$a_3 Pr {}^{CF}D_\tau^\beta \theta(\xi, \tau) = \lambda_{hnf} \frac{\partial^2 \theta(\xi, \tau)}{\partial \xi^2} + Q\theta(\xi, \tau), \tag{8.10}$$

where ${}^{CF}D_\tau^\alpha v(\xi, \tau)$ and ${}^{CF}D_\tau^\beta \theta(\xi, \tau)$ are the Caputo–Fabrizio fractional operators of fractional order α and β. Equations (8.9) and (8.10) are the Caputo–Fabrizio generalized form of Eqs. (8.5) and (8.6), while the initial and boundary conditions will remain the same as in Eqs. (8.7) and (8.8). The Caputo–Fabrizio fractional operator is defined by [2]

$${}^{CF}D_t^\delta f(t) = \frac{N(\delta)}{1 - \delta} \int_0^t \exp\left(-\frac{\delta(t - \tau)}{1 - \delta}\right) \frac{\partial f(\tau)}{\partial \tau} \mathrm{d}\tau, 0 < \delta < 1, \tag{8.11}$$

which is the convolution product of the function $\frac{N(\delta)}{1-\delta} \exp\left(-\frac{\delta t}{1-\delta}\right)$ and $f(t)'$ of fractional order δ. In this study, the following two properties of Caputo–Fabrizio fractional operator will be utilized:

1. **Property 1:** According to Losanda and Nieto [29], $N(\delta)$ is the normalization function such that

$$N(1) = N(0) = 1. \tag{8.12}$$

2. **Property 2:** Taking into consideration Eq. (8.12), the Laplace transform of Eq. (8.11) yields to

$$L\left\{{}^{CF}D_t^\delta f(t)\right\}(q) = \frac{q\overline{f}(q) - f(0)}{(1 - \delta) q + \delta}, 0 < \delta < 1, \tag{8.13}$$

Such that

$$\lim_{\delta\to1}\left[L\left\{{}^{CF}D_t^{\delta}f(t)\right\}(q)\right]=\lim_{\delta\to1}\left\{\frac{q\overline{f}(q)-f(0)}{(1-\delta)q+\delta}\right\}$$
$$=q\overline{f}(q)-f(0)=L\left\{\frac{\partial f(t)}{\partial t}\right\},\qquad(8.14)$$

where $\overline{f}(q)$ is the Laplace transform of $f(t)$ and $f(0)$ is the initial value of the function.

8.4 Solution of the Problem

To solve Eqs. (8.9) and (8.10), the Laplace transform method $L\{f(t)\}(q)=\overline{f}(q)=\int_0^{\infty}f(t)\,e^{-qt}\mathrm{d}t$ will be applied by using the corresponding initial and boundary conditions from Eqs. (8.7) and (8.8) to develop exact analytical solutions for velocity and temperature profiles.

8.4.1 Solutions of the Energy Equation

Applying the Laplace transform to Eq. (8.10) by keeping in mind the definition and properties of Caputo–Fabrizio fractional operator defined in Eqs. (8.11)–(8.14) and using the corresponding initial condition, Eq. (8.7) yields

$$a_3Pr\frac{q\theta(\xi,\tau)-\theta(\xi,0)}{(1-\beta)q+\beta}=\lambda_{hnf}\frac{d^2\overline{\theta}(\xi,q)}{\partial\xi^2}+Q\overline{\theta}(\xi,q),0<\beta<1.\qquad(8.15)$$

After further simplification, Eq. (8.15) gives

$$\frac{\mathrm{d}^2\overline{\theta}(\xi,\mathrm{q})}{\partial\xi^2}-\frac{b_4q-b_1b_3}{q+b_1}=0,0<\beta<1,\qquad(8.16)$$

with transformed boundary conditions:

$$\left.\begin{matrix}\overline{v}(0,q)=0,\overline{\theta}(0,q)=0, & \text{for}\,q>0,\\ \overline{v}(1,q)=0,\overline{\theta}(1,q)=\frac{1}{q}, & \text{for}\,q>0,\end{matrix}\right\},\qquad(8.17)$$

where

$$b_0=\frac{1}{1-\beta},\;b_1=b_0\beta,\;b_2=\frac{a_3Pr}{\lambda_{hnf}},\;b_3=\frac{Q}{\lambda_{hnf}},\;b_4=b_0b_2-b_3.$$

The exact solution of Eq. (8.16) using the corresponding boundary conditions from Eq. (8.17) is given by

$$\overline{\theta}(\xi,q)=\frac{1}{q}\left(\sinh\xi\sqrt{\frac{b_4q-b_1b_3}{q+b_1}}\right)\left(\sinh\sqrt{\frac{b_4q-b_1b_3}{q+b_1}}\right)^{-1},0<\beta<1.\qquad(8.18)$$

Equation (8.18) represents the solutions of the energy equation in the Laplace transformed domain. To invert the Laplace transform, this equation can be written in a more suitable and simplified form as

$$\overline{\theta}(\xi,q)=\sum_{n=0}^{\infty}\left(\frac{1}{q}e^{-(1+2n-\xi)\sqrt{\frac{b_4q-b_1b_3}{q+b_1}}}-\frac{1}{q}e^{-(1+2n+\xi)\sqrt{\frac{b_4q-b_1b_3}{q+b_1}}}\right),$$
$$0<\beta<1. \tag{8.19}$$

Let us consider

$$\overline{\theta}(\xi,q)=\overline{\theta}_1(\xi,q)-\overline{\theta}_2(\xi,q),0<\beta<1, \tag{8.20}$$

where

$$\overline{\theta}_1(\xi,q)=\sum_{n=0}^{\infty}\frac{1}{q}e^{-(1+2n-\xi)\sqrt{\frac{b_4q-b_1b_3}{q+b_1}}},0<\beta<1, \tag{8.21}$$

$$\overline{\theta}_2(\xi,q)=\sum_{n=0}^{\infty}\frac{1}{q}e^{-(1+2n+\xi)\sqrt{\frac{b_4q-b_1b_3}{q+b_1}}},0<\beta<1. \tag{8.22}$$

Upon taking the inverse Laplace transform, Eq. (8.20) yields

$$\theta(\xi,\tau)=\theta_1(\xi,\tau)-\theta_2(\xi,\tau),0<\beta<1. \tag{8.23}$$

To derive the functions $\theta_1(\xi,\tau)$ and $\theta_2(\xi,\tau)$, the compound formula for Laplace inversion is used. The function $\overline{\Phi}(\xi,q)$ is being chosen as

$$\overline{\Phi}(\xi,q)=e^{-(1+2\mathrm{n}-\xi)\sqrt{\frac{b_4q-b_1b_3}{q+b_1}}}=e^{(1+2\mathrm{n}-\xi)\sqrt{W_1(q)}}. \tag{8.24}$$

According to Khan [30], the inverse Laplace transform of the functions $\overline{\Phi}(\xi,q)$ can be obtained as

$$\Phi(\xi,\tau)=L^{-1}\left\{\overline{\Phi}(\xi,q)\right\}=\int_0^{\infty}f((1+2n-\xi),u)\,g(u,\tau)\,\mathrm{d}\tau, \tag{8.25}$$

where

$$f((1+2n-\xi),u)=\frac{(1+2n-\xi)}{2u\sqrt{\pi u}}e^{-\frac{(1+2n-\xi)}{4u}}, \tag{8.26}$$

$$g(u,\tau)=e^{-b_4u}\delta(\tau)-e^{-b_4u}\sqrt{\frac{pu}{\tau}}I_1\sqrt{pu\tau}e^{-b_1\tau}, \tag{8.27}$$

and

$$p=-b_1(b_3+b_4),$$

where $I_{1(.)}$ is the modified Bessel function. The values of functions $f\left((1+2n-\xi),u\right)$ and $g\left(u,\tau\right)$, defined in Eqs. (8.26) and (8.27), are used in Eq. (8.25) to yield the following simplified form:

$$\Phi\left(\xi,\tau\right)=e^{-(1+2n-\xi)\sqrt{b_4}}\delta\left(\tau\right)-\frac{\left(1+2n-\xi\right)\sqrt{p}}{2\sqrt{\pi\tau}}e^{-b_1\tau}$$
$$\int_0^{\infty}\frac{1}{u}e^{-\frac{(1+2n-\xi)^2}{4u}-b_4u}I_1\sqrt{pu\tau}du. \tag{8.28}$$

To evaluate the function $\theta_1\left(\xi,\tau\right)$, we need to find the convolutions product of $L^{-1}\left\{\frac{1}{q}\right\}=1$ and the function $\overline{\Phi}\left(\xi,q\right)$, which yield to

$$\theta_1\left(\xi,\tau\right)=\sum_{n=0}^{\infty}\left(e^{-(1+2n-\xi)\sqrt{b_4}}-\int_0^{\infty}\int_0^{\tau}\frac{\left(1+2n-\xi\right)\sqrt{p}}{2\sqrt{\pi s}}e^{-b_1 s}\right.$$
$$\left.\times\frac{1}{u}e^{-\frac{(1+2n-\xi)^2}{4u}-b_4u}I_1\sqrt{pus}\mathrm{d}u\mathrm{d}s\right). \tag{8.29}$$

Similarly, the function $\theta_2\left(\xi,\tau\right)$ is given by

$$\theta_1\left(\xi,\tau\right)=\sum_{n=0}^{\infty}\left(e^{-(1+2n+\xi)\sqrt{b_4}}-\int_0^{\infty}\int_0^{\tau}\frac{\left(1+2n+\xi\right)\sqrt{p}}{2\sqrt{\pi s}}e^{-b_1 s}\right.$$
$$\left.\times\frac{1}{u}e^{-\frac{(1+2n+\xi)^2}{4u}-b_4u}I_1\sqrt{pus}\mathrm{d}u\mathrm{d}s\right), \tag{8.30}$$

To reduce the solutions obtained in Eq. (8.23) to classical or local form, Eq. (8.14) is used. For $\beta\to 1$, Eq. (8.15) is reduced to the following form

$$\frac{\mathrm{d}^2\overline{\theta}\left(\xi,\mathrm{q}\right)}{\partial\xi^2}-\left(b_2q+b_2\right)\overline{\theta}\left(\xi,q\right)=0, \tag{8.31}$$

with the solutions in the Laplace transform domain being given by

$$\overline{\theta}\left(\xi,q\right)=\frac{1}{q}\frac{\sinh\xi\sqrt{b_2q-b_3}}{\sinh\sqrt{b_2q-b_3}}. \tag{8.32}$$

After further simplification, Eq. (8.32) takes the following form

$$\overline{\theta}\left(\xi,q\right)=\sum_{n=0}^{\infty}\left(\frac{1}{q}e^{-(1+2n-\xi)\sqrt{b_2q-b_3}}-\frac{1}{q}e^{-(1+2n+\xi)\sqrt{b_2q-b_3}}\right),\quad\beta=1. \tag{8.33}$$

Taking the inverse Laplace transform, Eq. (8.33) gives the following local solutions for the temperature profile:

$$\theta\left(\xi,\tau\right)=A_1\left(\xi,\tau\right)-A_2\left(\xi,\tau\right),\quad\beta=1, \tag{8.34}$$

where

$$A_1(\xi,\tau)=\frac{1}{2}\sum_{n=0}^{\infty}\left(e^{-\frac{(1+2n-\xi)}{\sqrt{b_2}}\sqrt{-\frac{b_3}{b_2}}}\operatorname{erfc}\left(\frac{(1+2n-\xi)}{2\sqrt{\tau}}-\sqrt{-\frac{b_3}{b_2}\tau}\right)\right.$$
$$\left.+e^{\frac{(1+2n-\xi)}{\sqrt{b_2}}\sqrt{-\frac{b_3}{b_2}}}\operatorname{erfc}\left(\frac{(1+2n-\xi)}{2\sqrt{\tau}}+\sqrt{-\frac{b_3}{b_2}\tau}\right)\right), \quad (8.35)$$

$$A_2(\xi,\tau)=\frac{1}{2}\sum_{n=0}^{\infty}\left(e^{-\frac{(1+2n+\xi)}{\sqrt{b_2}}\sqrt{-\frac{b_3}{b_2}}}\operatorname{erfc}\left(\frac{(1+2n-\xi)}{2\sqrt{\tau}}-\sqrt{-\frac{b_3}{b_2}\tau}\right)\right.$$
$$\left.+e^{\frac{(1+2n+\xi)}{\sqrt{b_2}}\sqrt{-\frac{b_3}{b_2}}}\operatorname{erfc}\left(\frac{(1+2n-\xi)}{2\sqrt{\tau}}+\sqrt{-\frac{b_3}{b_2}\tau}\right)\right). \quad (8.36)$$

8.4.2 Solution of Momentum Equation

Applying the Laplace transform on Eq. (8.9) using the corresponding initial condition from Eq. (8.5) yield

$$\frac{a_0}{a_1}\frac{q\overline{v}(\xi,q)-v(\xi,0)}{(1-\alpha)q+\alpha}+\frac{a_0}{a_1}\beta_b\overline{v}(\xi,q)=\frac{\mathrm{d}^2\overline{v}(\xi,q)}{\mathrm{d}\xi^2}+\frac{a_2}{a_1}Gr\overline{\theta}(\xi,\tau). \quad (8.37)$$

After further simplification of Eq. (8.37), it gives

$$\frac{\mathrm{d}^2\overline{v}(\xi,q)}{\mathrm{d}\xi^2}-\frac{d_5q+d_1d_3}{q+d_1}\overline{v}(\xi,q)=-d_4\overline{\theta}(\xi,q), \quad (8.38)$$

where

$$d_0=\frac{1}{1-\alpha},\ d_1=\alpha d_0,\ d_2=\frac{a_0}{a_1},\ d_3=d_2\beta_b,\ d_4=\frac{a_2}{a_1}Gr,\ d_5=d_0d_2+d_3.$$

In the Laplace transform domain, the exact solution of Eq. (8.38) is given by

$$\overline{v}(\xi,q)=\frac{(q+b_1)(q+b_1)}{(d_6q^2+d_7q+d_8)q}\frac{\sinh\xi\sqrt{\frac{d_5q+d_1d_3}{q+d_1}}}{\sinh\sqrt{\frac{d_5q+d_1d_3}{q+d_1}}}$$
$$-\frac{(q+b_1)(q+b_1)}{(d_6q^2+d_7q+d_8)q}\frac{\sinh\xi\sqrt{\frac{b_4q-b_1b_3}{q+b_1}}}{\sinh\sqrt{\frac{b_4q-b_1b_3}{q+b_1}}}, \quad (8.39)$$

where

$$W_1(q)=\frac{d_5q+d_1d_3}{q+d_1},\ d_6=b_4-d_5,\ d_7$$
$$=b_4d_1-b_1(b_3+d_5),\ d_8=-b_1d_1(b_3+d_3).$$

In order to find the inverse Laplace transform, Eq. (8.39) can be written in a more convenient and simplified form as

$$\overline{v}(\xi,q) = \frac{(q+b_1)(q+b_1)}{(d_6q^2+d_7q+d_8)\,q}\sum_{n=0}^{\infty}\left(e^{-(1+2n-\xi)\sqrt{\frac{d_5q+d_1d_3}{q+d_1}}} - e^{-(1+2n+\xi)\sqrt{\frac{d_5q+d_1d_3}{q+d_1}}}\right) - \frac{(q+b_1)(q+b_1)}{(d_6q^2+d_7q+d_8)\,q}$$
$$\times\sum_{n=0}^{\infty}\left(e^{-(1+2n-\xi)\sqrt{\frac{b_4q-b_1b_3}{q+b_1}}} - e^{-(1+2n+\xi)\sqrt{\frac{b_4q-b_1b_3}{q+b_1}}}\right),$$
$$0<\alpha,\beta<1. \tag{8.40}$$

Consider

$$\overline{v}(\xi,q) = \overline{v}_1(\xi,q)\times\{\overline{v}_2(\xi,q)-\overline{v}_3(\xi,q)\} - \overline{v}_1(\xi,q)\times\{\overline{v}_4(\xi,q)-\overline{v}_5(\xi,q)\},\, 0<\alpha,\beta<1, \tag{8.41}$$

where

$$\overline{v}_1(\xi,q) = \frac{(q+b_1)(q+b_1)}{(d_6q^2+d_7q+d_8)}, \tag{8.42}$$

$$\overline{v}_2(\xi,q) = \frac{1}{q}\sum_{n=0}^{\infty}e^{-(1+2n-\xi)\sqrt{\frac{d_5q+d_1d_3}{q+d_1}}}, \tag{8.43}$$

$$\overline{v}_3(\xi,q) = \frac{1}{q}\sum_{n=0}^{\infty}e^{-(1+2n+\xi)\sqrt{\frac{d_5q+d_1d_3}{q+d_1}}}, \tag{8.44}$$

$$\overline{v}_4(\xi,q) = \frac{1}{q}\sum_{n=0}^{\infty}e^{-(1+2n-\xi)\sqrt{\frac{b_4q-b_1b_3}{q+b_1}}}, \tag{8.45}$$

$$\overline{v}_5(\xi,q) = \frac{1}{q}\sum_{n=0}^{\infty}e^{-(1+2n+\xi)\sqrt{\frac{b_4q-b_1b_3}{q+b_1}}}. \tag{8.46}$$

The inverse Laplace transform yields

$$v(\xi,\tau) = v_1(\xi,\tau)*\{v_2(\xi,\tau)-v_3(\xi,\tau)\} - v_1(\xi,\tau)*\{v_4(\xi,\tau)-v_5(\xi,\tau)\},$$
$$0<\alpha,\beta<1, \tag{8.47}$$

where* represents convolutions product, and the terms $v_2(\xi,\tau)$, $v_3(\xi,\tau)$, $v_4(\xi,\tau)$ and $v_5(\xi,\tau)$ are given by

$$v_2(\xi,\tau)=\sum_{n=0}^{\infty}\left\{e^{-(1+2n-\xi)\sqrt{d_5}}\delta(\tau)-\frac{(1+2n-\xi)\sqrt{p_2}}{2\sqrt{\pi\tau}}e^{-d_1\tau}\right.$$
$$\left.\int_0^{\infty}\frac{1}{u}e^{-\frac{(1+2n-\xi)^2}{4u}-b_4u}I_1\sqrt{p_2u\tau}\mathrm{d}u.\right\}, \tag{8.48}$$

$$v_3(\xi,\tau)=\sum_{n=0}^{\infty}\left\{e^{-(1+2n+\xi)\sqrt{d_5}}\delta(\tau)-\frac{(1+2n+\xi)\sqrt{p_2}}{2\sqrt{\pi\tau}}e^{-d_1\tau}\right.$$
$$\left.\int_0^{\infty}\frac{1}{u}e^{-\frac{(1+2n+\xi)^2}{4u}-b_4u}I_1\sqrt{p_2u\tau}\mathrm{d}u.\right\}, \tag{8.49}$$

$$v_4(\xi,\tau)=\sum_{n=0}^{\infty}\left\{e^{-(1+2n-\xi)\sqrt{b_4}}\delta(\tau)-\frac{(1+2n-\xi)\sqrt{p_2}}{2\sqrt{\pi\tau}}e^{-b_1\tau}\right.$$
$$\left.\int_0^{\infty}\frac{1}{u}e^{-\frac{(1+2n-\xi)^2}{4u}-b_4u}I_1\sqrt{p_2u\tau}\mathrm{d}u\right\}, \tag{8.50}$$

$$v_5(\xi,\tau)=\sum_{n=0}^{\infty}\left\{e^{-(1+2n+\xi)\sqrt{b_4}}\delta(\tau)-\frac{(1+2n+\xi)\sqrt{p_2}}{2\sqrt{\pi\tau}}e^{-b_1\tau}\right.$$
$$\left.\int_0^{\infty}\frac{1}{u}e^{-\frac{(1+2n+\xi)^2}{4u}-b_4u}I_1\sqrt{p_2u\tau}\mathrm{d}u\right\}, \tag{8.51}$$

where

$$p_2=\frac{d_1d_3-d_5d_1}{d_1^2}.$$

The term $v_1(\xi,\tau)$ is numerically obtained using Zakian's algorithm. In the literature, it is proven that the Zakian's algorithm is a stable way for the inverse Laplace transform, because truncated error for five multiple terms is negligible [31]. For the velocity profile, the local solutions can be recovered by making $\alpha,\beta\to 1$ in Eq. (8.37), which leads to the following solutions in the Laplace transform domain:

$$\overline{v}(\xi,q)=\frac{d_4}{(b_2-d_3)q-(b_3+d_3)}\frac{1}{q}\frac{\sinh\xi\sqrt{d_2q+d_3}}{\sinh\sqrt{d_2q+d_3}}$$
$$-\frac{d_4}{(b_2-d_3)q-(b_3+d_3)}\frac{1}{q}\frac{\sinh\xi\sqrt{b_2q-b_3}}{\sinh\sqrt{b_2q-b_{33}}}, \tag{8.52}$$

with the following simplified form:

$$\overline{v}(\xi,q)=\frac{d_9}{q^2+qd_{10}}\sum_{n=0}^{\infty}\left(e^{-(1+2n-\xi)\sqrt{d_2q+d_3}}-e^{-(1+2n+\xi)\sqrt{d_2q+d_3}}\right)$$

$$- \frac{d_9}{q^2 + q d_{10}} \sum_{n=0}^{\infty} \left(e^{-(1+2n-\xi)\sqrt{b_2 q - b_3}} - e^{-(1+2n+\xi)\sqrt{b_2 q - b_3}} \right),$$

$$\alpha, \beta = 1, \tag{8.53}$$

where

$$d_9 = \frac{d_4}{b_2 - d_2}, \, d_{10} = \frac{b_3 + d_3}{b_2 - d_2}.$$

The inverse Laplace transform of Eq. (8.53) yields to

$$v(\xi, \tau) = B_1(\xi, \tau) - B_2(\xi, \tau) - B_3(\xi, \tau) + B_4(\xi, \tau) \quad \alpha, \beta = 1, \tag{8.54}$$

where

$$B_1(\xi, \tau) = \frac{d_{10}}{2} \sum_{n=0}^{\infty} \int_0^{\tau} e^{-d_7(\tau - s)} \left(e^{-\frac{(1+2n-\xi)}{\sqrt{b_2}}\sqrt{\frac{d_3}{d_2}}} \operatorname{erfc}\left(\frac{(1+2n-\xi)}{2\sqrt{s}} - \sqrt{\frac{d_3}{d_2} s} \right) \right.$$
$$\left. + e^{\frac{(1+2n-\xi)}{\sqrt{b_2}}\sqrt{-\frac{b_3}{b_2}}} \operatorname{erfc}\left(\frac{(1+2n-\xi)}{2\sqrt{s}} + \sqrt{\frac{d_3}{d_2} s} \right) \right) ds, \tag{8.55}$$

$$B_2(\xi, \tau) = \frac{d_{10}}{2} \sum_{n=0}^{\infty} \int_0^{\tau} e^{-d_7(\tau - s)} \left(e^{-\frac{(1+2n+\xi)}{\sqrt{b_2}}\sqrt{\frac{d_3}{d_2}}} \operatorname{erfc}\left(\frac{(1+2n-\xi)}{2\sqrt{s}} - \sqrt{\frac{d_3}{d_2} s} \right) \right.$$
$$\left. + e^{\frac{(1+2n+\xi)}{\sqrt{b_2}}\sqrt{-\frac{d_3}{d_2}}} \operatorname{erfc}\left(\frac{(1+2n-\xi)}{2\sqrt{s}} + \sqrt{\frac{d_3}{d_2} s} \right) \right) ds, \tag{8.56}$$

$$B_3(\xi, \tau) = \frac{d_{10}}{2} \sum_{n=0}^{\infty} \int_0^{\tau} e^{-d_7(\tau - s)} \left(e^{-\frac{(1+2n-\xi)}{\sqrt{b_2}}\sqrt{-\frac{b_3}{b_2}}} \operatorname{erfc} \right.$$
$$\times \left(\frac{(1+2n-\xi)}{2\sqrt{s}} - \sqrt{-\frac{b_3}{b_2} s} \right) + e^{\frac{(1+2n-\xi)}{\sqrt{b_2}}\sqrt{-\frac{b_3}{b_2}}} \operatorname{erfc}$$
$$\left. \times \left(\frac{(1+2n-\xi)}{2\sqrt{s}} + \sqrt{-\frac{b_3}{b_2} s} \right) \right) ds, \tag{8.57}$$

$$B_4(\xi, \tau) = \frac{d_{10}}{2} \sum_{n=0}^{\infty} \int_0^{\tau} e^{-d_7(\tau - s)} \left(e^{-\frac{(1+2n+\xi)}{\sqrt{b_2}}\sqrt{-\frac{b_3}{b_2}}} \operatorname{erfc} \right.$$
$$\times \left(\frac{(1+2n-\xi)}{2\sqrt{s}} - \sqrt{-\frac{b_3}{b_2} s} \right) + e^{\frac{(1+2n+\xi)}{\sqrt{b_2}}\sqrt{-\frac{b_3}{b_2}}} \operatorname{erfc}$$
$$\left. \times \left(\frac{(1+2n-\xi)}{2\sqrt{s}} + \sqrt{-\frac{b_3}{b_2} s} \right) \right) ds. \tag{8.58}$$

8.5 Results and Discussion

In this chapter, the idea of fractional derivative is used for the generalization of the free convection flow hybrid nanofluid. The governing equations of Brinkman-type fluid along with energy equation is fractionalized using Caputo–Fabrizio fractional derivative. The fractional partial differential equations (PDEs) are more general and are known as master PDEs. The momentum and energy equations are solved analytically using the Laplace transform technique. The obtained results are displayed in various graphs to study the influence of pertinent corresponding parameters such as fractional parameters α and β and the volume fraction of hybrid nanofluid.

Figures 8.1a and b and 8.2a and b depict the impact of the fractional parameter 'α' and β on velocity and temperature profiles. From these figures, it is noticed that both the velocity and temperature profiles show the same

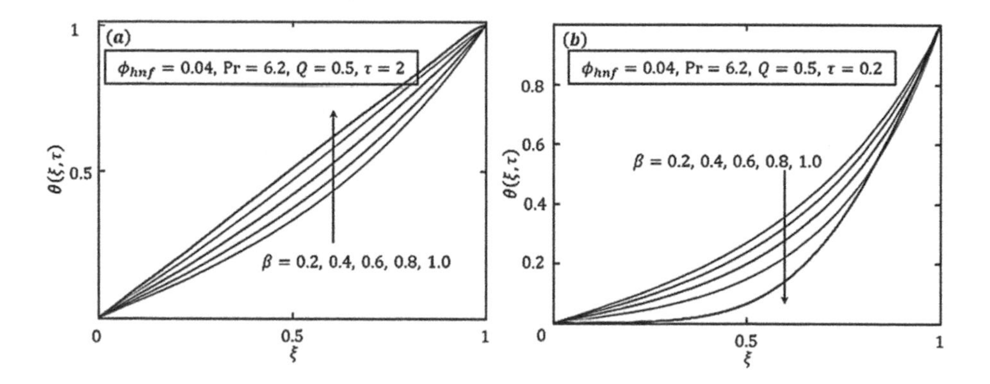

FIGURE 8.1
Variation in temperature profile against ξ due to β.

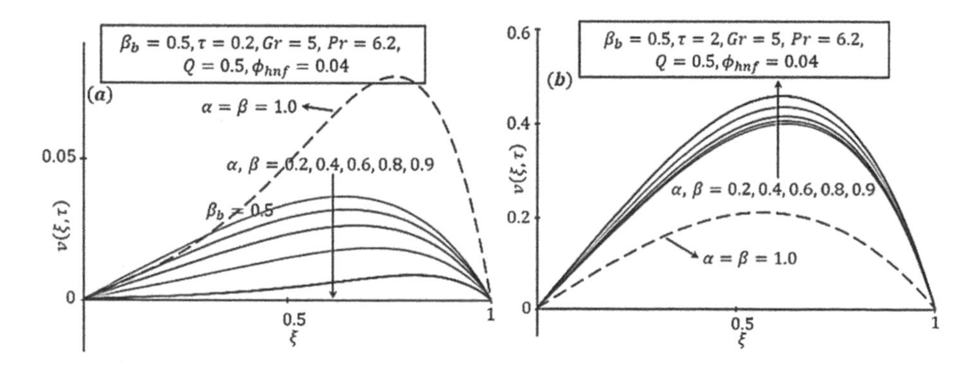

FIGURE 8.2
Variation in velocity profile against ξ due to α and β.

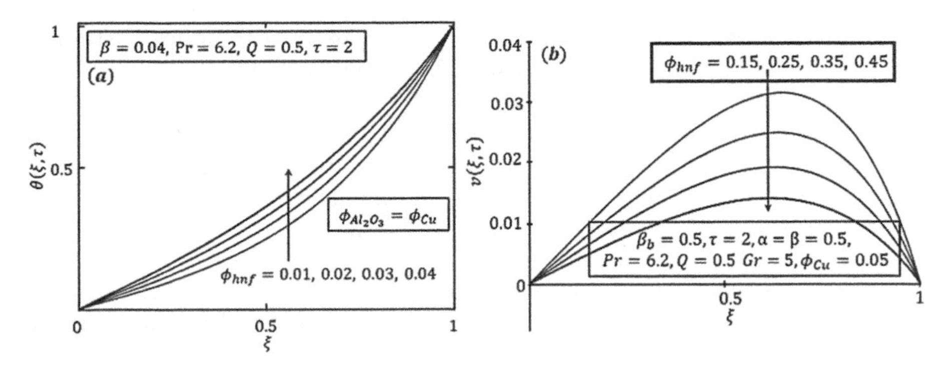

FIGURE 8.3
Variation in temperature and velocity profile against ξ due to ϕ_{hnf}.

trend for variations in fractional parameters. The velocity and temperature profiles exhibited an increasing behaviour for increasing values of α and β for a longer time. When α, β are increased, the thickness of thermal and momentum boundary layers are increased and became the thickest at $\alpha, \beta = 1$, which correspond to the increasing performance of velocity and temperature profiles. This manner of the fractional parameter is same here for velocity and temperature profiles, as reported by Azhar and Khan et al. [18,32] for fractional nanofluid using Caputo–Fabrizio fractional derivatives. But this effect reverses for a shorter time in case of fractional velocity and temperature distributions.

The influence of ϕ_{hnf} on the velocity and temperature profiles is studied in Figure 8.3a and b. The trend of velocity and temperature profiles is opposite to each other. The hybrid nanofluid velocity decreases with increasing ϕ_{hnf}. This can be physically justifying as the hybrid nanofluid became more viscous with increasing ϕ_{hnf}, which leads to a decrease in the nanofluid velocity. Nevertheless, the temperature profile increases with an increase in ϕ_{hnf} when the temperature is less than 180°C. This is because thermal conductivity enhances with enhancement of ϕ_{hnf}, and the hybrid nanofluid conducts more heat as the resultant heat transfer increases, which leads to an increase in the temperature profile.

8.6 Concluding Remarks

In this chapter, the idea of free convection is generalized using Caputo–Fabrizio fractional derivative. The natural convection flow of hybrid nanofluid in two vertical infinite parallel plates is studied. Exact analytical solutions are developed for temperature and velocity profiles via the Laplace transform technique. The effects of various pertinent parameters are numerically studied through graphs and discussed physically. The major key points extracted from this study are as follows:

- The velocity and temperature profiles show an increasing behaviour for increasing values α and β and are most dominant for $\alpha, \beta = 1$ for a larger time. But this effect reverses for a shorter time.
- The fractional velocity and temperature are more general. Hence, the numerical values for $v(\xi, \tau)$ and $\theta(\xi, \tau)$ can be calculated for any value of α and β, such that $0 < \alpha,\ \beta < 1$.
- The temperature distribution shows a very similar variation for different shapes of hybrid nanoparticles, and so the density of the nanoparticles is a significant factor when compared with thermal conductivity.
- The velocity profile decreases with increasing values of ϕ_{hnf}, but this effect is opposite in case of the temperature profile.

Acknowledgment

The authors acknowledge the Ministry of Eduction (MOE) and Research Management Centre-UTM, Universiti Teknologi Malaysia, UTM, for the financial support through vote numbers 5F004 and 07G70, 07G72, 07G76 and 07G77 for this research.

References

[1] Podlubny I. *Fractional Differential Equations: An Introduction to Fractional Derivatives, Fractional Differential Equations, to Methods of Their Solution and Some of Their Applications.* Vol. 198. Elsevier, San Diego; 1998.

[2] Caputo M, Fabrizio M. A new definition of fractional derivative without singular kernel. *Progress in Fractional Differentiation and Applications.* 2015;1(2):1–13.

[3] Atangana A, Baleanu D. New fractional derivatives with nonlocal and non-singular kernel: theory and application to heat transfer model. arXiv preprint arXiv:160203408. 2016.

[4] Baleanu D, Agheli B, Darzi R. An optimal method for approximating the delay differential equations of noninteger order. *Advances in Difference Equations*. 2018;2018(1):284.

[5] Yavuz M, Ozdemir N, Baskonus HM. Solutions of partial differential equations using the fractional operator involving Mittag-Leffler kernel. *The European Physical Journal Plus*. 2018;133(6):215.

[6] Saqib M, Khan I, Shafie S. Natural convection channel flow of CMC-based CNTs nanofluid. *The European Physical Journal Plus*. 2018;133(12):549.

[7] Yavuz M, Özdemir N. A different approach to the European option pricing model with new fractional operator. *Mathematical Modelling of Natural Phenomena*. 2018;13(1):12.

[8] Saqib M, Khan I, Shafie S. Application of Atangana–Baleanu fractional derivative to MHD channel flow of CMC-based-CNT's nanofluid through a porous medium. *Chaos, Solitons & Fractals*. 2018;116:79–85.

[9] Saqib M, Khan I, Shafie S. Application of fractional differential equations to heat transfer in hybrid nanofluid: modeling and solution via integral transforms. *Advances in Difference Equations*. 2019;2019(1):52.

[10] Zafar AA, Fetecau C. Flow over an infinite plate of a viscous fluid with non-integer order derivative without singular kernel. *Alexandria Engineering Journal*. 2016;55(3):2789–2796.

[11] Makris N, Dargush GF, Constantinou MC. Dynamic analysis of generalized viscoelastic fluids. *Journal of Engineering Mechanics*. 1993;119(8):1663–1679.

[12] Alkahtani BST, Atangana A. Modeling the potential energy field caused by mass density distribution with Eton approach. *Open Physics*. 2016;14(1):106–113.

[13] Vieru D, Fetecau C, Fetecau C. Time-fractional free convection flow near a vertical plate with Newtonian heating and mass diffusion. *Thermal Science*. 2015;19(suppl. 1):85–98.

[14] Abro KA, Khan I, Gómez-Aguilar JF. A mathematical analysis of a circular pipe in rate type fluid via Hankel transform. *The European Physical Journal Plus*. 2018;133(10):397.

[15] Jain S. Numerical analysis for the fractional diffusion and fractional Buckmaster equation by the two-step Laplace Adam-Bashforth method. *The European Physical Journal Plus*. 2018;133(1):19.

[16] Yavuz M, Ozdemir N. Numerical inverse Laplace homotopy technique for fractional heat equations. *Thermal Science*. 2018;22(1):185–194.

[17] Saqib M, Khan I, Shafie S. New direction of Atangana–Baleanu fractional derivative with mittag-leffler kernel for non-newtonian channel flow. In José Francisco Gómez, Lizeth Torres, Ricardo Fabricio Escobar (eds.), *Fractional Derivatives with Mittag-Leffler Kernel.* Springer, Polish Academy of Sciences, Warsaw; 2019. pp. 253–268.

[18] Azhar WA, Vieru D, Fetecau C. Free convection flow of some fractional nanofluids over a moving vertical plate with uniform heat flux and heat source. *Physics of Fluids.* 2017;29(8):082001.

[19] Jain, S., & Atangana, A. Analysis of lassa hemorrhagic fever model with non-local and non-singular fractional derivatives. *International Journal of Biomathematics.* 2018;11(08):1850100.

[20] Yavuz M, Özdemir N. European vanilla option pricing model of fractional order without singular kernel. *Fractal and Fractional.* 2018;2(1):3.

[21] Wang X-Q, Mujumdar AS. Heat transfer characteristics of nanofluids: a review. *International Journal of Thermal Sciences.* 2007;46(1):1–19.

[22] Sheikholeslami M, Shamlooei M, Moradi R. Numerical simulation for heat transfer intensification of nanofluid in a porous curved enclosure considering shape effect of Fe_3O_4 nanoparticles. *Chemical Engineering and Processing: Process Intensification.* 2018;124:71–82.

[23] Hussanan A, Salleh MZ, Khan I, Shafie, S. Convection heat transfer in micropolar nanofluids with oxide nanoparticles in water, kerosene and engine oil. *Journal of Molecular Liquids.* 2017;229:482–488.

[24] Bhattad A, Sarkar J, Ghosh P. Discrete phase numerical model and experimental study of hybrid nanofluid heat transfer and pressure drop in plate heat exchanger. *International Communications in Heat and Mass Transfer.* 2018;91:262–273.

[25] Hussain S, Ahmed SE, Akbar T. Entropy generation analysis in MHD mixed convection of hybrid nanofluid in an open cavity with a horizontal channel containing an adiabatic obstacle. *International Journal of Heat and Mass Transfer.* 2017;114:1054–1066.

[26] Farooq U, Afridi M, Qasim M, Lu D. Transpiration and viscous dissipation effects on entropy generation in hybrid nanofluid flow over a nonlinear radially stretching disk. *Entropy.* 2018;20(9):668.

[27] Jan SAA, Ali F, Sheikh NA, Khan I, Saqib M, Gohar M. Engine oil based generalized brinkmantype nano-liquid with molybdenum disulphide nanoparticles of spherical shape: Atangana–Baleanu fractional model. *Numerical Methods for Partial Differential Equations.* 2018;34(5): 1472–1488.

[28] Khan I, Saqib M, Ali F. Application of the modern trend of fractional differentiation to the MHD flow of a generalized Casson fluid in a microchannel: modelling and solution. *The European Physical Journal Plus.* 2018;133(7):262.

[29] Losada J, Nieto JJ. Properties of a new fractional derivative without singular kernel. *Progress in Fractional Differentiation and Applications.* 2015;1(2):87–92.

[30] Khan I. A note on exact solutions for the unsteady free convection flow of a Jeffrey fluid. *Zeitschrift für Naturforschung A.* 2015;70(6):397–401.

[31] Zakian V, Littlewood RK. Numerical inversion of Laplace transforms by weighted least-squares approximation. *The Computer Journal.* 1973;16(1):66–68.

[32] Khan I, Saqib M, Ali F. Application of time-fractional derivatives with non-singular kernel to the generalized convective flow of Casson fluid in a microchannel with constant walls temperature. *The European Physical Journal Special Topics.* 2017;226(16–18):3791–3802.

9

Collation Analysis of Fractional Moisture Content Based Model in Unsaturated Zone Using q-homotopy Analysis Method

Ritu Agarwal and Mahaveer Prasad Yadav
Malaviya National Institute of Technology

Ravi P. Agarwal
Texas A&M University - Kingsville

CONTENTS

9.1 Introduction

In unsaturated medium, the water flow through soils and pore spaces are not completely filled with flowing groundwater because the substances changes with respect to time. The process of recharging groundwater by spreading has great importance to hydrologists, agriculturalists and engineers. This type of water flow helps to encounter underground disposal of seepage and wastewater intrusion, described by a non-linear partial differential equation.

Moisture content of a soil is the quantity of water contained in soil. Here, some assumptions are made that the area of groundwater recharge is a large basin, geological formation of this basin is bounded by stiff boundaries and on the lower side by a dense layer of watertable and flow is assumed to be vertically downward in an unsaturated medium. Here, we assume that the diffusion coefficient is equal to its mean value over the entire area and the permeability coefficient is a continuous linear map to the moisture content in soil.

Soil is an important part in the hydrological cycle process. Underground formation of soil is divided into two zones: the saturated zone in which the entire void space is occupied by water and an unsaturated zone in which the void space contains air as well as water. In the unsaturated zone, moisture content is zero, while in the saturated zone, moisture content has a value of 1.

Analytic solutions for the groundwater movement through the unsaturated medium are analyzed by solving Richard's equation in [6,25,26,28,30]. Several numerical models were developed using finite element method, finite difference method and integrated finite difference method in [10,11,13–15,22,23]. In these models, most of the numerical methods are focused on improving the existing methods. Recently, Prasad et al. [27] developed a numerical model using finite element method to simulate moisture flow through an unsaturated medium. Agarwal et al. [2–4, 31] found analytical and numerical solutions of many groundwater flow problems. Nguyen et al. [24] proposed a numerical flow approach based on isogeometric analysis in an unsaturated medium. Shahrokhadi et al. [29] introduced a head-based solution to the Richard's equation based on isogeometric analysis for unsaturated flow in porous media. Abid [1] developed a finite volume model to predict the moisture-based form in unsaturated porous media. In an unsaturated zone, the equation of continuity is given as

$$\frac{\partial}{\partial t}(\rho_s\theta) = -\nabla.M, \tag{9.1}$$

where ρ_s is the density of the unsaturated zone, θ is the moisture content and M is the mass flux of the moisture content. Velocity of the groundwater in porous media is given by Darcy law as

$$\bar{V} = -k\nabla\phi, \tag{9.2}$$

where $\nabla\phi$ is the gradient of the moisture potential and k is the aqueous conductivity. From Eqs. (9.1) and (9.2), we obtain

$$\frac{\partial}{\partial t}(\rho_s\theta) = -\nabla.(\rho k\nabla\phi), \tag{9.3}$$

where ρ is the flux density and $M = \rho\bar{V}$. Here, we assume that when the flow is in the downward direction, then Eq. (9.3) becomes

$$\rho_s\frac{\partial\theta}{\partial t} = \frac{\partial}{\partial z}\left(\rho k\frac{\partial\psi}{\partial z}\right) - \frac{\partial}{\partial z}(\rho kg), \tag{9.4}$$

where ψ is the capillary pressure potential, g is the gravitational constant and $\phi = \psi - gz$. The z-axis and gravity are both in the positive direction. Here, we consider that the θ and ψ are connected by a single-valued function; thus, we can write Eq. (9.4) as

$$\frac{\partial\theta}{\partial t} = \frac{\partial}{\partial z}\left(D\frac{\partial\theta}{\partial z}\right) - \frac{\rho g}{\rho_s}\frac{\partial k}{\partial z}, \tag{9.5}$$

where $D = \frac{\rho}{\rho_s} k \frac{\partial \psi}{\partial z}$. Replacing D by mean value D_a and assuming $k = k_0 \theta^p$, $p = 1, 2$, in Eq. (9.5), we have

$$\frac{\partial \theta}{\partial t} = D_a \frac{\partial^2 \theta}{\partial z^2} - \frac{\rho g}{\rho_s} k_0 p \theta^{p-1} \frac{\partial \theta}{\partial z}. \tag{9.6}$$

Let the watertable be situated at depth L from ground and set non-dimensional variables as

$$\frac{z}{L} = \xi, \ \frac{tD_a}{L^2} = T, \ \beta = \frac{\rho g}{\rho_s} \frac{k_0}{D_a}, \ 0 \le \xi \le 1, \ 0 \le T \le 1, \ 0 \le \beta \le 1. \tag{9.7}$$

Equation (9.6) becomes

$$\frac{\partial \theta(\xi, T)}{\partial T} = \frac{\partial^2 \theta(\xi, T)}{\partial \xi^2} - \beta p \theta^{p-1}(\xi, T) \frac{\partial \theta(\xi, T)}{\partial \xi}, \tag{9.8}$$

with initial and boundary conditions

$$\theta(\xi, 0) = \theta_0(\xi) = f(\xi), \tag{9.9}$$

where ξ is the penetration depth, T is the time and β is the flow parameter. The moisture content at the layer ($z = 0$) in the unsaturated porous media is very small, and at the watertable ($z = L$) the moisture content of soil will be fully saturated. So it is suitable to choose the boundary condition as

$$\lim_{\xi \to 0} \theta(\xi, T) \to 0, \ \lim_{\xi \to 1^-} \theta(\xi, T) \to 1. \tag{9.10}$$

9.2 Mathematical Preliminaries

Fractional differential operators with non-singular kernel have captured the minds of several researchers because of its applicability to almost all fields of science, engineering and technology with different scales. Riemann–Liouville and Caputo fractional differential operators could not model the complex real-world problems as they have a singular kernel in their integral, whereas the fractional differential operators with non-singular kernel have removed this issue, as they are able to incorporate the effect of memory and long-range dependence into the mathematical formulation.

Caputo fractional derivative [8, Eq. 5, p. 530] of a function $f : [a, b] \to \mathbb{R}$ of order α and $t > a$ is given by

$${}_a^C D_t^{\alpha} f(t) = \frac{1}{\Gamma(\alpha - 1)} \int_a^t \frac{f'(\tau) d\tau}{(t - \tau)^{\alpha}}, \ 0 < \alpha < 1, \tag{9.11}$$

where $\Gamma(.)$ is the Gamma function.

Laplace transform of the Caputo fractional derivative is given by Caputo [7] and Kilbas et al. [16, Eq. 2.4.62, p. 98] as

$$\mathcal{L}\left[{}_a^C D_t^\alpha f(t); s\right] = s^\alpha F(s) - \sum_{k=0}^{n-1} s^{\alpha-k-1} f^{(k)}(0),\ (n-1) < \alpha \leq n, \tag{9.12}$$

where $Re(s) > 0$ and $Re(\alpha) > 0$ and $\mathcal{L}[f(t); s] = F(s)$ is the Laplace transform of $f(t)$.

Let $f \in H^{'}(a, b),\ b > a$ then, the definition of Caputo fractional derivative is given by Caputo and Fabrizio [9] as

$${}_a^{CF} D_t^\alpha f(t) = \frac{M(\alpha)}{1-\alpha} \int_a^t f^{'}(\tau) \exp\left(-\alpha \frac{t-\tau}{1-\alpha}\right) d\tau,\ a < t < b, \tag{9.13}$$

where $\alpha \in [0, 1]$, $M(\alpha)$ is a normalized function s.t. $M(0) = M(1) = 1$.

When $f(t)$ is constant, the fractional derivative is zero, according to the definition (9.11) and (9.13) which matches with the boundary conditions of a problem. Caputo fractional derivative has a singular kernel, while Caputo-Fabrizio derivative does not have singularity at $t = \tau$.

Laplace transform of the Caputo–Fabrizio fractional derivative is given by Caputo and Fabrizio [9] as

$$\mathcal{L}\left[{}_0^{CF} D_t^\alpha f(t); s\right] = M(\alpha) \frac{sF(s) - f(0)}{s + \alpha(1-s)},\ 0 < \alpha \leq 1, \tag{9.14}$$

if exists.

Recently, the Atangana–Baleanu fractional derivative [5] is introduced, which is very useful in modelling the complex real-world problems more accurately because of its wide applicability. It covers very important properties such as Markovian and non-Markovian, power law, stretched exponential and Brownian motion, Gaussian and non-Gaussian compared with other fractional derivative operators.

Let $f \in H^{'}(a, b),\ b > a,\ \alpha \in [0, 1]$ then, the definition of the Atangana–Baleanu fractional derivative in Caputo sense [5] is given as

$${}_a^{ABC} D_t^\alpha f(t) = \frac{M(\alpha)}{1-\alpha} \int_a^t f^{'}(\tau) E_\alpha \left(-\alpha \frac{(t-\tau)^\alpha}{1-\alpha}\right) d\tau, \tag{9.15}$$

where $M(\alpha)$ is a normalized function s.t. $M(0) = M(1) = 1$, and is defined as

$$M(\alpha) = 1 - \alpha + \frac{\alpha}{\Gamma(\alpha)}, \tag{9.16}$$

and $E_\nu(.)$ is the Mittag-Leffler function [21] given in series form as

$$E_\nu(z) = \sum_{n=0}^{\infty} \frac{z^n}{\Gamma(n\nu + 1)},\ z \in \mathbb{C},\ \nu \in (0, 1]. \tag{9.17}$$

The kernel in Eq. (9.15) is a combination of both exponential law and power law. Laplace transform of the Atangana–Baleanu fractional derivative of a function $f(t)$ is given as

$$\mathcal{L}\left[{}_{0}^{ABC}D_t^{(\alpha)} f(t); s\right] = M(\alpha)\frac{s^{\alpha}F(s) - s^{\alpha-1}f(0)}{s^{\alpha}(1-\alpha)+\alpha},\ 0 < \alpha \leq 1, \tag{9.18}$$

if exists.

Remark 9.1 *These three fractional derivatives, i.e. Caputo, Caputo–Fabrizio and Atangana–Baleanu, agree with integer order derivative for integer values of* α, *i.e.* $\alpha \in \mathbb{N}$.

9.3 Fractional Moisture Content Based Model

Here, we use the q-homotopy analysis method (q-HAM) to solve fractional moisture content based model. The solution of this new model is obtained by replacing time derivatives by ${}_{0}^{C}D_t^{\alpha}$, ${}_{0}^{CF}D_t^{\alpha}$*and* ${}_{0}^{ABC}D_t^{\alpha}$, respectively, in Eqs. (9.8)–(9.9), as

$${}_{0}^{C}D_t^{\alpha}\theta(\xi,T) + \beta p\theta^{p-1}(\xi,T)\frac{\partial\theta(\xi,T)}{\partial\xi} - \frac{\partial^2\theta(\xi,T)}{\partial\xi^2} = 0, \tag{9.19}$$

$${}_{0}^{CF}D_t^{\alpha}\theta(\xi,T) + \beta p\theta^{p-1}(\xi,T)\frac{\partial\theta(\xi,T)}{\partial\xi} - \frac{\partial^2\theta(\xi,T)}{\partial\xi^2} = 0, \tag{9.20}$$

$${}_{0}^{ABC}D_t^{\alpha}\theta(\xi,T) + \beta p\theta^{p-1}(\xi,T)\frac{\partial\theta(\xi,T)}{\partial\xi} - \frac{\partial^2\theta(\xi,T)}{\partial\xi^2} = 0, \tag{9.21}$$

where $0 < \alpha \leq 1$, $\xi \in \mathbb{R}$, $T > 0$ and ${}_{0}^{C}D_t^{\alpha}$, ${}_{0}^{CF}D_t^{\alpha}$, ${}_{0}^{ABC}D_t^{\alpha}$ are the Caputo, Caputo–Fabrizio and Atangana–Baleanu fractional derivatives of order α given in Eqs. (9.11), (9.13) and (9.15), respectively.

Applying Laplace transform on time variable T w.r.t. s, Eqs. (9.19), (9.20) and (9.21) become

$$s^{\alpha}\Theta(\xi,s) - \frac{\theta(\xi,0)}{s^{1-\alpha}} + \mathcal{L}\left[\beta p\theta^{p-1}(\xi,T)\frac{\partial\theta(\xi,T)}{\partial\xi} - \frac{\partial^2\theta(\xi,T)}{\partial\xi^2}; s\right] = 0, \tag{9.22}$$

$$M(\alpha)\frac{s\Theta(\xi,s) - \theta(\xi,0)}{s+\alpha(1-s)} + \mathcal{L}\left[\beta p\theta^{p-1}(\xi,T)\frac{\partial\theta(\xi,T)}{\partial\xi} - \frac{\partial^2\theta(\xi,T)}{\partial\xi^2}; s\right] = 0, \tag{9.23}$$

$$M(\alpha)\frac{s^{\alpha}\Theta(\xi,s) - s^{\alpha-1}\theta(\xi,0)}{s^{\alpha}(1-\alpha)+\alpha} + \mathcal{L}\left[\beta p\theta^{p-1}(\xi,T)\frac{\partial\theta(\xi,T)}{\partial\xi} - \frac{\partial^2\theta(\xi,T)}{\partial\xi^2}; s\right] = 0, \tag{9.24}$$

where $\Theta(\xi, s)$ is the Laplace transform of $\theta(\xi, T)$. On simplifying Eqs. (9.22), (9.23) and (9.24), we have

$$\Theta(\xi, s) - \frac{1}{s}\theta(\xi, 0) + u(.)\mathcal{L}\left[\beta p\theta^{p-1}(\xi, T)\frac{\partial\theta(\xi, T)}{\partial\xi} - \frac{\partial^2\theta(\xi, T)}{\partial\xi^2}; s\right] = 0, \tag{9.25}$$

where $u(.)$ is given by $u(C) = \frac{1}{s^\alpha}$, $u(CF) = \dfrac{s + \alpha(1 - s)}{sM(\alpha)}$ and $u(ABC)$ $= \dfrac{\alpha(-1 + s^{-\alpha}) + 1}{M(\alpha)}$, respectively.

Now we define a non-linear operator given in [12] as

$$\mathcal{N}[\sigma] = \mathcal{L}[\sigma; s] - \frac{\theta(\xi, 0)}{s}\left(1 - \frac{c_m}{n}\right) + u(.)\mathcal{L}\left[\beta p\sigma^{p-1}\frac{\partial\sigma}{\partial\xi} - \frac{\partial^2\sigma}{\partial\xi^2}; s\right], \tag{9.26}$$

where $q \in [0, \frac{1}{n}]$ is the embedding parameter, $n \in \mathbb{N}$ and $\sigma(\xi, T; q) \equiv \sigma_q$ is the supplementary function and

$$c_m = \begin{cases} 0 & \text{if } m \leq 1, \\ n & \text{if } m > 1. \end{cases} \tag{9.27}$$

Liao [17–20] constructed a zero-order deformation equation as

$$(1 - nq)\mathcal{L}[\sigma - \theta_0(\xi, T); s] = qhH(\xi, s)\mathcal{N}[\sigma], \tag{9.28}$$

where h is the non-zero auxiliary parameter and $H(\xi, s)$ is a non-zero auxiliary function, respectively, and $\theta_0(\xi, T)$ is the initial guess for $\theta(\xi, T)$. Now, when $q = 0$ and $q = \frac{1}{n}$, we have

$$\sigma(\xi, T; 0) \equiv \sigma_0 = \theta_0(\xi, T), \quad \sigma\left(\xi, T; \frac{1}{n}\right) \equiv \sigma_{\frac{1}{n}} = \theta(\xi, T), \tag{9.29}$$

respectively. So from Eq. (9.29), we can say that as q increases from 0 to $\frac{1}{n}$, the solution $\sigma(\xi, T; q)$ varies from the initial guess $\theta_0(\xi, T)$ to the solution $\theta(\xi, T)$. Here, we have the freedom to choose $\theta_0(\xi, T)$, h, $H(\xi, s)$ so that the solution $\sigma(\xi, T; q)$ of Eq. (9.28) exists for $q \in [0, \frac{1}{n}]$. Now, expanding the supplementary function $\sigma(\xi, T; q)$ in Taylor series form about q, we have

$$\sigma = \theta_0(\xi, T) + \sum_{m=1}^{\infty} \theta_m(\xi, T)q^m, \tag{9.30}$$

where

$$\theta_m(\xi, T) = \frac{1}{m!}\frac{\partial^m\sigma}{\partial q^m}\bigg|_{q=0}. \tag{9.31}$$

If $\theta_0(\xi, T)$, h and $H(\xi, s)$ are chosen properly, then the series (9.30) converges at $q = \frac{1}{n}$ and we have

$$\theta(\xi, T) = \theta_0(\xi, T) + \sum_{m=1}^{\infty} \theta_m(\xi, T) \left(\frac{1}{n}\right)^m, \tag{9.32}$$

Now differentiating Eq. (9.28) m times w.r.t. q and dividing by $m!$ at $q = 0$, we have the mth-order deformation equation

$$\mathcal{L}[\theta_m(\xi, T) - c_m \theta_{m-1}(\xi, T); s] = hH(\xi, s)\mathcal{R}_m(\vec{\theta}_{m-1}(\xi, T)), \tag{9.33}$$

where

$$\mathcal{R}_m(\vec{\theta}_{m-1}) = \frac{1}{(m-1)!} \frac{\partial^{m-1}\mathcal{N}(\sigma)}{\partial q^{m-1}}\bigg|_{q=0}, \tag{9.34}$$

and $\vec{\theta}_m(\xi, T)$ is the vector defined as

$$\vec{\theta}_m(\xi, T) = (\theta_0(\xi, T),\ \theta_1(\xi, T),\ \theta_2(\xi, T), \ldots, \theta_m(\xi, T)). \tag{9.35}$$

On taking inverse Laplace transformation of Eq. (9.33), we get the following recursive relation as

$$\theta_m(\xi, T) = c_m \theta_{m-1}(\xi, T) + h\mathcal{L}^{-1}[H(\xi, s)\mathcal{R}_m(\vec{\theta}_{m-1}(\xi, T)); T], \tag{9.36}$$

where

$$\begin{aligned}\mathcal{R}_m(\vec{\theta}_{m-1}(\xi, T)) &= \mathcal{L}[\theta_{m-1}(\xi, T); s] - \frac{\theta(\xi, 0)}{s}\left(1 - \frac{c_m}{n}\right) \\ &+ u(.)\mathcal{L}\left[p\beta \sum_{i=0}^{m-1} \theta_i^{p-1}(\xi, T) \frac{\partial \theta_{m-1-i}(\xi, T)}{\partial \xi} - \frac{\partial^2 \theta_{m-1}(\xi, T)}{\partial \xi^2}; s\right].\end{aligned} \tag{9.37}$$

Rest of the approximations is obtained by following the same procedure; therefore, the approximation solution of Eqs. (9.19)–(9.21) is obtained as

$$\theta(\xi, T) = \theta_0(\xi, T) + \sum_{i=1}^{m} \frac{\theta_i(\xi, T)}{n^i}. \tag{9.38}$$

It should be accented that $\theta_m(\xi, T)$ for $m \geq 1$ is governed by the linear equation (9.36), and due to the factor $\left(\frac{1}{n}\right)^m$, more chances of faster convergences may occur than the standard homotopy analysis method. In this method, we have greater freedom to choose the auxiliary parameter h, auxiliary function $H(\xi, T)$ and the initial guesses. The q-HAM is a simple, yet powerful analytic–numeric method for solving fractional differential equations. By this method, we can adjust and control the convergence region of solution series by choosing proper values for auxiliary parameter h and auxiliary function $H(\xi, T)$. If we take $n = 1$ in Eq. (9.28), standard homotopy analysis method can be obtained.

9.4 Applications

Here, as applications, we have taken different initial conditions for $\theta(\xi,0) = f(\xi)$ and taking auxiliary function $H(\xi,T) = 1$. Considering $f(\xi) = \xi^2$, we successively obtain the approximation solutions of Eqs. (9.19)–(9.21) as

$$\theta_0(\xi,T) = \theta(\xi,0) = \xi^2, \tag{9.39}$$

$$\theta_1^C(\xi,T) = h\left(2p\beta\xi^{2p-1} - 2\right)\frac{T^\alpha}{\Gamma(\alpha+1)}, \tag{9.40}$$

$$\begin{aligned}\theta_2^C(\xi,T) &= nh\left(2p\beta\xi^{2p-1} - 2\right)\frac{T^\alpha}{\Gamma(\alpha+1)} + h\frac{2h^2(p\beta\xi^{2p-1}-1)T^\alpha}{\Gamma(\alpha+1)}\\ &+ \frac{2h^2(p\beta)^2(2p-1)\xi^{4p-4}T^{2\alpha}}{\Gamma(2\alpha+1)}\\ &+ 2h\xi p\beta\left(\frac{2h}{\Gamma(\alpha+1)}(p\beta\xi^{2p-1}-1)\right)^{p-1}\frac{\Gamma(\alpha(p-1)+1)T^{\alpha p}}{\Gamma(\alpha p+1)}\\ &- 2hp\beta(2p-1)(2p-2)\xi^{2p-3}\frac{T^{2\alpha}}{\Gamma(2\alpha+1)}.\end{aligned} \tag{9.41}$$

$$\theta_1^{CF}(\xi,T) = 2h(p\beta\xi^{2p-1}-1)(1-\alpha+t\alpha), \tag{9.42}$$

$$\begin{aligned}\theta_2^{CF}(\xi,T) &= 2nh(p\beta\xi^{2p-1}-1)(1-\alpha+T\alpha) + 2h^2(p\beta\xi^{2p-1}-1)(1-\alpha+T\alpha)\\ &+ \left((1-\alpha)(1-\alpha+2T\alpha) + \frac{\alpha^2T^2}{2}\right)(2h^2(p\beta)^2(2p-1)\xi^{4p-4}\\ &- 2h^2p\beta(2p-1)(2p-2)\xi^{2p-3}) - 2hp\beta\xi\mathcal{L}^{-1}\left[\left(1-\alpha+\frac{\alpha}{s}\right)\right.\\ &\left.\times\mathcal{L}\left[\left(2h(1-\alpha+T\alpha)(p\beta\xi^{2p-1}-1)\right)^{p-1}\right]\right].\end{aligned} \tag{9.43}$$

$$\theta_1^{AB}(\xi,T) = 2h(p\beta\xi^{2p-1}-1)\left(1-\alpha+\alpha\frac{T^\alpha}{\Gamma(\alpha+1)}\right), \tag{9.44}$$

$$\begin{aligned}\theta_2^{AB}(\xi,T) &= 2nh(p\beta\xi^{2p-1}-1)\left(1-\alpha+\alpha\frac{T^\alpha}{\Gamma(\alpha+1)}\right) + 2h^2(p\beta\xi^{2p-1}-1)\\ &\times\left(1-\alpha+\alpha\frac{T^\alpha}{\Gamma(\alpha+1)}\right) + \left((1-\alpha)\left(1-\alpha+\alpha\frac{T^\alpha}{\Gamma(\alpha+1)}\right)\right.\end{aligned}$$

$$+ \frac{\alpha^2 T^{2\alpha}}{\Gamma(2\alpha+1)}\Big) (2h^2(p\beta)^2(2p-1)\xi^{4p-4} - 2h^2 p\beta(2p-1)(2p-2)$$

$$\times \xi^{2p-3}) - 2hp\beta\xi\mathcal{L}^{-1}\left[\left(1-\alpha+\frac{\alpha}{s^\alpha}\right)\right.$$

$$\left.\times \mathcal{L}\left[\left(2h\left(1-\alpha+\alpha\frac{T^\alpha}{\Gamma(\alpha+1)}\right)(p\beta\xi^{2p-1}-1)\right)^{p-1}\right]\right]. \quad (9.45)$$

Considering $f(\xi) = e^{-\xi}$, we successively obtain the approximation solutions of Eqs. (9.19)–(9.21) as

$$\theta_0(\xi, T) = \theta(\xi, 0) = e^{-\xi}, \quad (9.46)$$

$$\theta_1^C(\xi, T) = -h\left(p\beta e^{-\xi p} + e^{-\xi}\right)\frac{T^\alpha}{\Gamma(\alpha+1)}, \quad (9.47)$$

$$\theta_2^C(\xi, T) = -nh\left(p\beta e^{-\xi p} + e^{-\xi}\right)\frac{T^\alpha}{\Gamma(\alpha+1)} + (hp^3\beta^2 e^{\xi-2\xi p}$$

$$+ (p\beta + hp^3\beta)e^{-\xi p} + he^{-\xi})\frac{T^{2\alpha}}{\Gamma(2\alpha+1)}$$

$$- p\beta e^{-\xi}h^{p-1}\left(\frac{(p\beta e^{-\xi p} - e^{-\xi})}{\Gamma(\alpha+1)}\right)^{p-1}\frac{\Gamma(\alpha(p-1)+1)T^{\alpha p}}{\Gamma(\alpha p+1)} \quad (9.48)$$

$$\theta_1^{CF}(\xi, T) = -h(p\beta e^{-\xi p} + e^{-\xi})(1-\alpha+T\alpha), \quad (9.49)$$

$$\theta_2^{CF}(\xi, T) = -nh(p\beta e^{-\xi p} + e^{-\xi})(1-\alpha+T\alpha) - h^2(p\beta e^{-\xi p} + e^{-\xi})(1-\alpha+T\alpha)$$

$$+ hp\beta(hp^2\beta e^{\xi-2\xi p} + h^2 e^{-\xi p})\left((1-\alpha)^2 + 2\alpha(1-\alpha)t + \frac{\alpha^2 T^2}{2}\right)$$

$$- 2hp\beta\xi\mathcal{L}^{-1}\left[\left(1-\alpha+\frac{\alpha}{s}\right)\mathcal{L}\left[(2h(1-\alpha+T\alpha)(p\beta\xi^{2p-1}-1))^{p-1}\right]\right]. \quad (9.50)$$

$$\theta_1^{AB}(\xi, T) = -h(p\beta e^{-\xi p} + e^{-\xi})\left(1-\alpha+\alpha\frac{T^\alpha}{\Gamma(\alpha+1)}\right) \quad (9.51)$$

$$\theta_2^{AB}(\xi, T) = -nh(p\beta e^{-\xi p} + e^{-\xi})\left(1-\alpha+\alpha\frac{T^\alpha}{\Gamma(\alpha+1)}\right)$$

$$- h(p\beta e^{-\xi p} + e^{-\xi})\left(1-\alpha+\alpha\frac{T^\alpha}{\Gamma(\alpha+1)}\right)$$

$$
\begin{aligned}
&+\mathcal{L}^{-1}\left[\left(1-\alpha+\frac{\alpha}{s^{\alpha}}\right)\mathcal{L}\left[h\beta pe^{\xi-\xi p}(p^2\beta e^{-\xi p}+e^{-\xi})\right.\right.\\
&\times\left(1-\alpha+\alpha\frac{T^{\alpha}}{\Gamma(\alpha+1)}\right)+h\left(\beta p^3e^{-\xi p}+e^{-\xi}\right)\\
&\times\left(1-\alpha+\alpha\frac{T^{\alpha}}{\Gamma(\alpha+1)}\right)+h^{p-1}e^{-\xi}\left(\beta pe^{-\xi p}+e^{-\xi}\right)^{p-1}\\
&\left.\left.\times\left(1-\alpha+\alpha\frac{T^{\alpha}}{\Gamma(\alpha+1)}\right)^{p-1}\right]\right]. \qquad (9.52)
\end{aligned}
$$

9.5 Numerical Simulation

In this section, we compare the approximation solutions of moisture content based model obtained by applying Caputo, Caputo–Fabrizio and Atangana–Baleanu fractional derivative operators, and the behaviour of moisture content w.r.t. various parameters has been observed, and a graphical representation of moisture content w.r.t. to time variable is presented. The figures are plotted via MATLAB®. Solution of Eq. (9.9) is plotted graphically for the Caputo, Caputo–Fabrizio and Atangana–Baleanu fractional derivative operators, respectively, for $\alpha = 0.50,\ 0.90$ and fixed value of the parameters $\beta = 3.5, h = 0.1, p = 1, \xi = 0.5$ for $f(x) = \xi^2$.

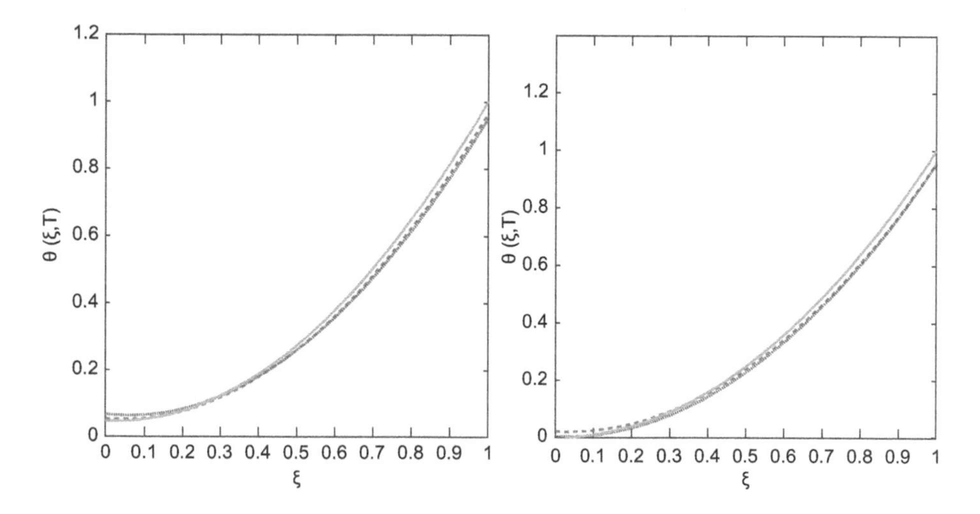

FIGURE 9.1
The q-HAM solutions of Eqs. (9.19)–(9.21) for Caputo (dashed line), Caputo–Fabrizio (dashed-dot–dashed line) and Atangana–Baleanu (solid line) fractional derivative operators, respectively, for $\alpha = 0.5$ and 0.9.

From Figure 9.1, we observe that moisture content is increased w.r.t. spatial variable ξ when groundwater is recharged by spreading in an unsaturated homogeneous porous media. In Figure 9.1 at $\alpha = 0.5$, solutions are close to each other and when α goes close to 1, approximate solutions coincide with each other. From Figures 9.2 and 9.3 we observe that moisture content increases w.r.t. time variable T. For lower values of fractional order α, we observe instability there. Figure 9.3 is plotted for a non-linear case ($p = 2$). The results show that the convergence region of series solutions obtained by q-HAM is increasing as q is decreased.

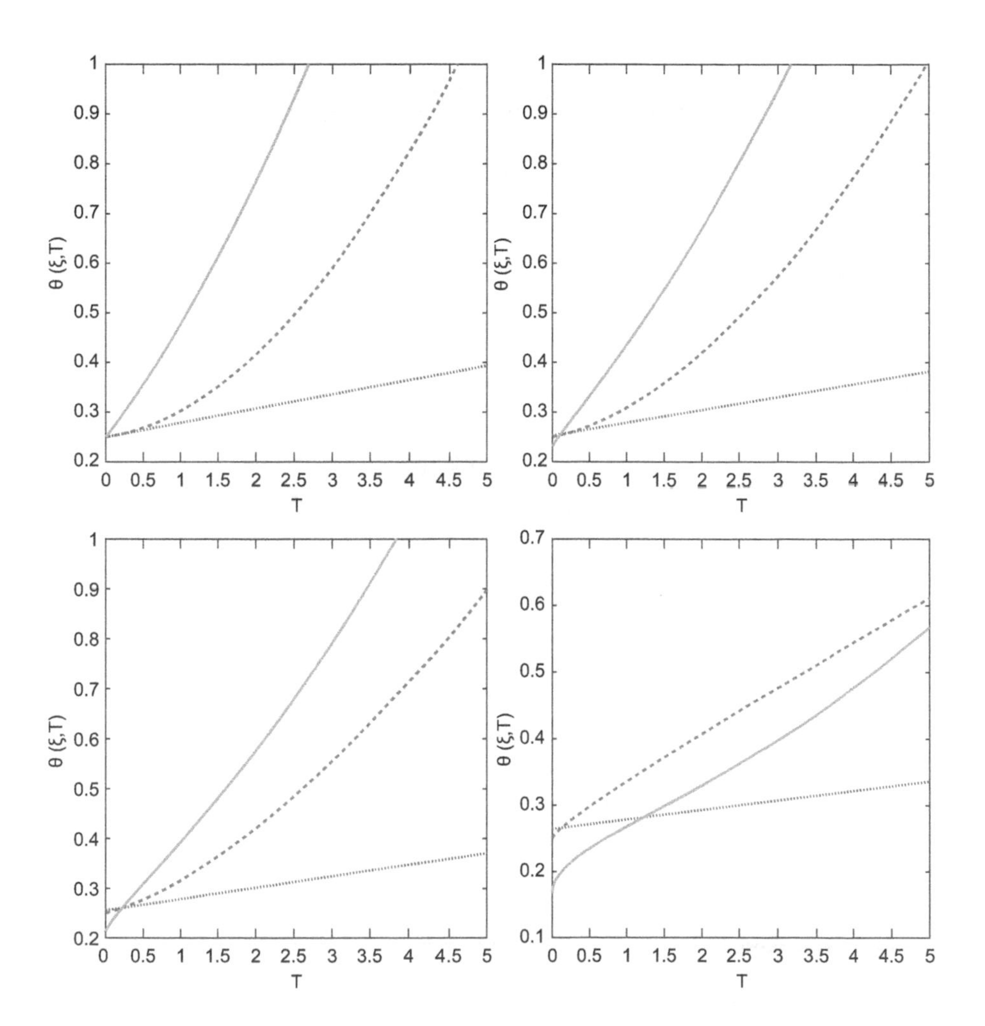

FIGURE 9.2
The q-HAM solutions of Eqs. (9.19)–(9.21) for Caputo (dashed line), Caputo–Fabrizio (dashed-dot–dashed line) and Atangana–Baleanu (solid line) fractional derivative operators, respectively, for $\alpha = 1$ and 0.9, 0.8 and 0.5.

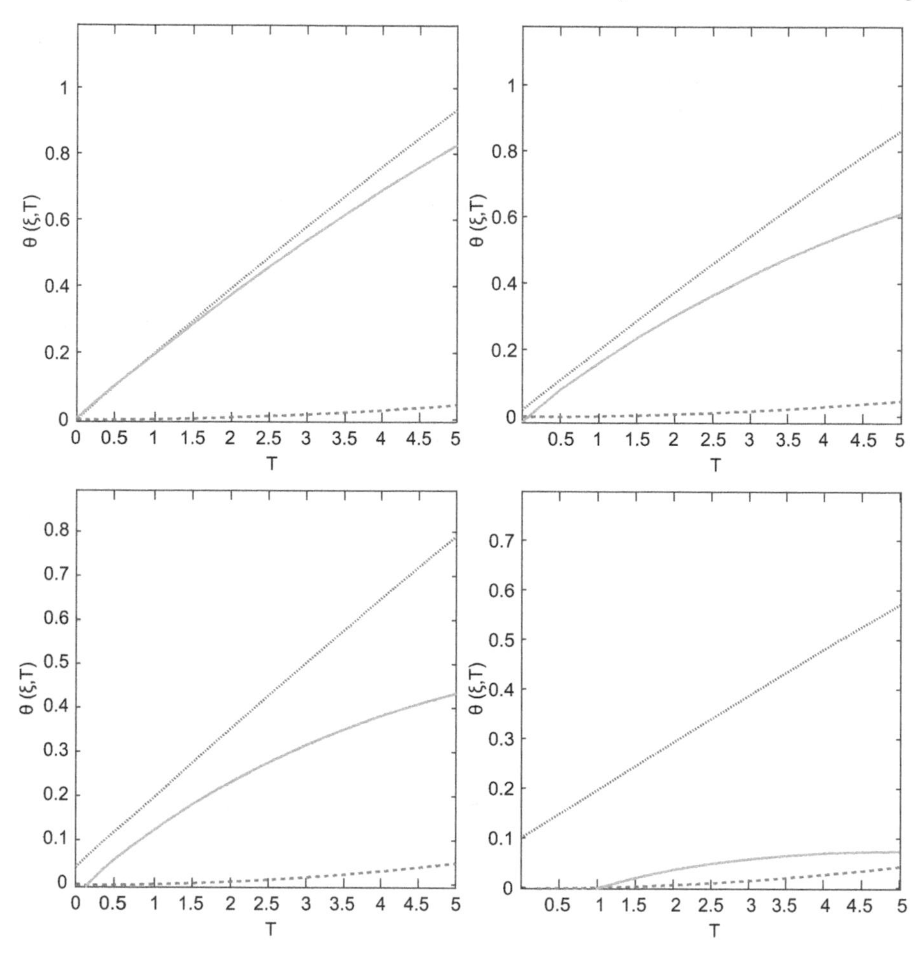

FIGURE 9.3
The q-HAM solutions of Eqs. (9.19)–(9.21) for Caputo (dashed line), Caputo–Fabrizio (dashed-dot–dashed line) and Atangana–Baleanu (solid line) fractional derivative operators, respectively, for non-linear case ($p = 2$) and $\alpha = 1,\ 0.9,\ 0.8$ and 0.5.

9.6 Conclusion

In this work, the mathematical model of the one-dimensional groundwater recharge by spreading of the water on the ground in the downward direction is developed, and the mathematical formation of the model leads to a non-linear partial differential equation that has been fractionalized using the

Caputo, Caputo–Fabrizio and Atangana–Baleanu fractional derivative operators. q-HAM is applied to obtain the solution of the governing equations and plotted graphically for linear and non-linear cases. The moisture content in unsaturated porous medium is varying w.r.t. time $T > 0$. The solution being in series form, its components can be easily computed as far as we like using any computational software. In general, this method produces a rapidly convergent series.

References

[1] Abid, M. B. (2014). Numerical simulation of two-dimensional unsaturated flow from a trickle irrigation source using the finite-volume method, *Journal of Irrigation and Drainage Engineering*, **141**, 06014005-1–8.

[2] Agarwal, R., Yadav, M. P. and Agarwal, R. P. (2019). Analytic solution of space time fractional advection dispersion equation with retardation for contaminant transport in porous media, *Progress in Fractional Differentiation and Applications*, **5**(4), 1–13.

[3] Agarwal, R., Yadav, M. P. and Agarwal, R. P. (2019). Analytic solution of time fractional Boussinesq equation for groundwater flow in unconfined aquifer, *Journal of Discontinuity, Nonlinearity and Complexity*, **8**(3), 341–352.

[4] Agarwal, R., Yadav, M. P., Agarwal, R. P. and Goyal, R. (2019). Analytic solution of fractional advection dispersion equation with decay for contaminant transport in porous media, *Matematicki Vesnik*, **71**, 5–15.

[5] Atangana, A. and Baleanu, D. (2016). New fractional derivatives with nonlocal and non-singular kernel: Theory and application to heat transfer model, *arXiv:1602.03408*.

[6] Broadbridge, P. and White, I. (1988). Constant rate rainfall infiltration: A versatile nonlinear model, 1. Analytical solution, *Water Resources Research*, **24**, 145–154.

[7] Caputo, M. (1969). *Elasticita e Dissipazione*, Zanichelli, Bologna.

[8] Caputo, M. (1967). Linear models of dissipation whose Q is almost frequency independent II, *Geophysical Journal of the Royal Astronomical Society*, **13**, 529–539.

[9] Caputo, M. and Fabrizio, M. (2015). A new definition of fractional derivative without singular kernel, *Progress in Fractional Differentiation and Application*, **2**, 73–85.

[10] Celia, M. A., Bouloutas, E. T. and Zarba, R. L. (1990). A general mass conservative numerical solution for the unsaturated flow equation, *Water Resources Research*, **26**, 1483–1496.

[11] Cooley, R. L. (1983). Some new procedures for numerical solution of variably saturated flow problems, *Water Resource Research*, **19**, 1271–1285.

[12] El-Tawil, M. A. and Huseen, S. N. (2012). The q-homotopy analysis method (q-ham), *International Journal of Applied Mathematics and Mechanics*, **8**, 51–75.

[13] Gottardi, G. and Venutelli, M. (1992). Moving finite element model for one dimensional infiltration in unsaturated soil, *Water Resourch Research*, **28**, 3259–3267.

[14] Hills, R. G., Porro, I., Hudson, D. B. and Wierenga, J. P. (1989). Modeling one dimensional infiltration into very dry soils. 1. Model development and evaluation, *Water Resource Research*, **25**, 1259–1269.

[15] Huyakorn, P. S., Springer, E. P., Guvansen, V. and Wordsworth, T. D. (1986). A three dimensional finite element model for simulating water flow in variably saturated porous media, *Water Resource Research*, **22**, 1790–1808.

[16] Kilbas, A. A., Srivastava, H. M., and Trujillo, J. J. (2006). *Theory and Applications of Fractional Differential Equations,* Elsevier, Amsterdam.

[17] Liao, S. J. (2005). Comparison between the homotopy analysis method and homotopy perturbation method, *Applied Mathematics and Computation*, **169**, 1186–1194.

[18] Liao, S. J. (2004). On the homotopy analysis method for nonlinear problems, *Applied Mathematics and Computation*, **147**, 499–513.

[19] Liao, S. J. (2003). *Beyond Perturbation: Introduction to the Homotopy Analysis Method*, Chapman and Hall/CRC Press, Boca Raton, FL.

[20] Liao, S. J. (1992). The proposed homotopy analysis technique for the solution of nonlinear problems, PhD thesis, Shanghai Jiao Tong University.

[21] Mittag-Leffler, G. M. (1903). Sur la nouvelle fonction $E_\alpha(x)$, *Comptes Rendus de l'Academie des Sciences Paris (Srie II),* **137**, 554–558.

[22] Narasimhan, T. N. and Witherspoon, P. A. (1977). Numerical model for saturated-unsaturated flow in deformable porous media. 1. Theory, *Water Resource Research*, **13**, 657–664.

[23] Neuman, S. P. (1973). Saturated-unsaturated seepage by finite elements, *Journal of Hydraulic Engineering*, **99**, 2233–2250.

[24] Nguyen, M. N., Bui, T. Q., Yu, T. and Hirose, S. (2014). Isogeometric analysis for unsaturated flow problems, *Computers and Geotechnics*, **62**, 257–267.

[25] Parlange, J. (1972). Theory of water movement in soils, 6. Effect of water depth over soil, *Soil Science*, **113**, 308–312.

[26] Philip, J. R. (1969). Theory of infiltration, *Advances in Hydroscience*, **5**, 215–296.

[27] Prasad, K. S. H., Kumar, M. S. M. and M Sekhar, M. (2001). Modelling flow through unsaturated zones: Sensitivity to unsaturated soil properties, *Sadhana*, **26**, 517–528.

[28] Richards, L. A. (1931). Capillary conduction of liquids through porous medium, *Physics*, **1**, 313–318.

[29] Shahrokhabadi, S., Vahedifard, F. and Bhatia, M. (2017). Head-based isogeometric analysis of transient flow in unsaturated soils, *Computers and Geotechnics*, **84**, 183–197.

[30] Warrick, A. W., Islas, A. and Lomen, D. O. (1991). An analytical solution to Richards equation for time varying infiltration, *Water Resources Research*, **27**, 763–766.

[31] Yadav, M. P. and Agarwal, R. (2019). Numerical investigation of fractional-fractal Boussinesq equation, *Chaos: An Interdisciplinary Journal of Nonlinear Science*, **29**, 013109.

10

Numerical Analysis of a Chaotic Model with Fractional Differential Operators: From Caputo to Atangana–Baleanu

Jyoti Mishra
Gyan Ganga Institute of Technology and Sciences

Abdon Atangana
University of the Free State

CONTENTS

10.1 Introduction

Fractional calculus is regarded as applied mathematics. The theories and properties of these fractional operators have generated a lot of interest and became an active subject of study in the last few decades. Recently, scientists, engineers and applied mathematicians have found the fractional calculus concept useful in various fields, such as quantitative biology, theology, electrochemistry, diffusion process, scattering theory, transport theory, elasticity, probability and potential theory. A great revolution has taken place in the field of fractional calculus in the last decades as researchers have discovered that these mathematical tools can capture better physical complex problems that could not be with classical differentiation [1–10]. Nevertheless, within the field of fractional differentiation, one can so far identify three classes. Here we refer to the fractional calculus based on power law, and this one has been used in several fields but is subject to some limitations, as it cannot portray dynamical system taking place in different layers, which means this operator cannot handle physical problems displaying crossover properties [10–13]. Additionally, this operator includes an artificial singularity that forces a non-singular problem to become singular [1–5]. The second is new and is case on the exponential function, and this version suffers from the fact that it cannot capture non-locality linked to time; however, this calculus is able to describe physical problem with crossover that includes a steady state [10–14]. The additional important properties of this operator are its ability to describe fading memory processes and statistical properties like Gaussian distribution [15–17]. This fractional calculus is in fashion nowadays, as it has been applied in several fields with great results. The last and the very recent one is that based on the generalized Mittag-Leffler kernel. This last one was suggested to solve the problem of non-locality in time associate to the Caputo–Fabrizio calculus [13–14]. This last one is more natural, since many physical problems can be captured with such function. Additionally, these operators can be used to capture statistical properties like Gaussian, random walk and Brownian motion. Very importantly, the operator possesses the ability to capture power law and stretched exponential law [11–14]. Also, being in fashion nowadays and have been applied in many fields of science, this operator has opened new doors of investigation, in modelling, theory and applications. One of the properties of fractional differential operator is perhaps the ability to accurately replicate and reveal some chaotic; however, due to the non-linearity of these problems, their analytical solutions are harder and sometimes even impossible to obtain. Therefore, to appreciate their physical behaviours, researchers rely on numerical methods, and in this chapter, we present a numerical method for a chaotic problem.

10.2 Basic Definitions of Fractional Calculus

In this section, we report a quick tour of some of the useful basic definitions of fractional calculus.

Definition 10.1 The topic of fractional differentiation is one of the burning topics nowadays. In all of the fields these days, the fractional operators are powerful mathematical tools to solve real-world problems. The Riemann and Liouville operator is the first operator given as follows

$$^{RL}D_t^{\alpha}[f(t)] = \frac{1}{\Gamma(1-\alpha)}\frac{d}{dt}\int_0^t (t-y)^{-\alpha} f(y)dy \tag{10.1}$$

The fractional Caputo-type derivatives are defined as

$$^{c}D_t^{\alpha}[f(t)] = \frac{1}{\Gamma(1-\alpha)}\int_0^t (t-y)^{-\alpha}\frac{d}{dt} f(y)dy \tag{10.2}$$

Definition 10.2 Caputo and Fabrizio recommended a new trend of fractional differentiation, and they used the exponential decay law instead of the power law. This style of fractional differentiation is defined as

$$^{CF}_{0}D_t^{\alpha}\left[f(t)\right] = \frac{M(\alpha)}{(1-\alpha)}\int_0^t \frac{d}{dy} f(y)\exp\left[-\frac{\alpha}{1-\alpha}(t-y)\right]dy, \quad 0<\alpha<1 \tag{10.3}$$

Modified version of Caputo and Fabrizio fractional differentiation is suggested by Goufo and Atangana, which is defined as follows:

$$^{CF}_{0}D_t^{\alpha}[f(t)] = \frac{M(\alpha)}{(1-\alpha)}\frac{d}{dt}\int_0^t f(y)\exp\left[-\frac{\alpha}{1-\alpha}(t-y)\right]dy, \quad 0<\alpha<1 \tag{10.4}$$

Definition 10.3 Atangana and Baleanu recommended a new kernel on the generalized Mittag-Leffler function, which is the most suitable function that was introduced to solve some problems of disc of convergence of power law. The function is well thought out as the king of fractional calculus and is more likely than power law. Their definitions are given as

$$^{ABC}_{0}D_t^{\alpha}[f(t)] = \frac{B(\alpha)}{(1-\alpha)}\int_0^t \frac{d}{dy} f(y)E_{\alpha}\left[-\frac{\alpha}{1-\alpha}(t-y)^{\alpha}\right]dy, \quad 0<\alpha<1 \tag{10.5}$$

$$^{ABR}_{0}D_t^{\alpha}\left[f(t)\right] = \frac{B(\alpha)}{(1-\alpha)}\frac{d}{dt}\int_0^t f(y)E_{\alpha}\left[-\frac{\alpha}{1-\alpha}(t-y)^{\alpha}\right]dy, \quad 0<\alpha<1 \tag{10.6}$$

10.3 New Numerical Scheme with Atangana–Baleanu Fractional Derivative

In this section, we construct a new numerical scheme for non-linear fractional differential equations with fractional derivative with non-local and non-singular kernel. To do this, we consider a modified Lorenz chaotic model.

In this paper, to examine the chaotic behaviour of Lorenz chaotic model with Atangana–Baleanu fractional derivative, Miranda and Stone proposed a modified Lorenz chaotic model.

The modified Lorenz chaotic model considered here is given by

$$\begin{cases} \dfrac{dx(t)}{dt} = \dfrac{1}{3} \times (-(a+1)x(t) + a - c + z(t), y(t)) + ((1-a).(x(t)^2 - y(t)^2)) \\ \qquad + (2.(a+c-z(t)) \times x(t) \times y(t)).\dfrac{1}{3 \times \sqrt{x(t)^2 + y(t)^2}} \\ \dfrac{dy(t)}{dt} = \dfrac{1}{3} \times ((c-a-z(t)) \times x(t) - (a+1)y(t) + (2 \times (a-1) \times x(t) \times y(t) \\ \qquad + (a+c-z(t)) \times (x(t)^2 - y(t)^2)) \times \dfrac{1}{3 \times \sqrt{x(t)^2 + y(t)^2}} \\ \dfrac{dz(t)}{dt} = \dfrac{1}{2} \times (3 \times x(t)^2 \times y(t) - y(t)^3) - bz(t) \end{cases} \tag{10.7}$$

In order to check the effect of Mittage-Leffler non-local and non-singular kernel into the definition of fractional derivative, the time derivative of Eq. (10.7) will be replaced by the time Atangan–Baleanu fractional derivative to obtain

$$\begin{cases} {}_0^{ABC}D_t^\alpha x(t) = \dfrac{1}{3} \times (-(a+1)x(t) + a - c + z(t), y(t)) + ((1-a) \\ \quad .(x(t)^2 - y(t)^2)) + (2.(a+c-z(t)) \times x(t) \times y(t)).\dfrac{1}{3 \times \sqrt{x(t)^2 + y(t)^2}} \\ {}_0^{ABC}D_t^\alpha y(t) = \dfrac{1}{3} \times ((c-a-z(t)) \times x(t) - (a+1)y(t) + (2 \times (a-1) \\ \quad \times x(t) \times y(t) + (a+c-z(t)) \times (x(t)^2 - y(t)^2)) \times \dfrac{1}{3 \times \sqrt{x(t)^2 + y(t)^2}} \\ {}_0^{ABC}D_t^\alpha z(t) \doteq \dfrac{1}{2} \times (3 \times x(t)^2 \times y(t) - y(t)^3) - bz(t) \end{cases} \tag{10.8}$$

For simplicity, we assume that

$$\begin{cases} {}_0^{ABC}D_t^\alpha x(t) = f_1(x, y, z, t) \\ {}_0^{ABC}D_t^\alpha y(t) = f_2(x, y, z, t) \\ {}_0^{ABC}D_t^\alpha z(t) = f_3(x, y, z, t) \end{cases} \tag{10.9}$$

Applying the Atangana–Baleanu fractional integral on the both sides of system (10.9), we obtain

$$\left.\begin{aligned} x(t) &= x_0(t) + \frac{1-\alpha}{B(\alpha)} F_1(x, y, z, t) + \frac{\alpha}{B(\alpha)\Gamma(\alpha)} \int_0^t F_1(x, y, z, \tau)(t-\tau)^{\alpha-1} d\tau \\ y(t) &= y_0(t) + \frac{1-\alpha}{B(\alpha)} F_2(x, y, z, t) + \frac{\alpha}{B(\alpha)\Gamma(\alpha)} \int_0^t F_2(x, y, z, \tau)(t-\tau)^{\alpha-1} d\tau \\ z(t) &= z_0(t) + \frac{1-\alpha}{B(\alpha)} F_3(x, y, z, t) + \frac{\alpha}{B(\alpha)\Gamma(\alpha)} \int_0^t F_3(x, y, z, \tau)(t-\tau)^{\alpha-1} d\tau \end{aligned}\right\} \tag{10.10}$$

At a given point $t_{n\,+\,1}$, $n = 0, 1, 2$, the earlier equation is reformulated as

$$\left.\begin{aligned} x(t) - x_0 &= \frac{1-\alpha}{AB(\alpha)} f_1(x, y, z, \tau) + \frac{\alpha}{AB(\alpha)\Gamma(\alpha)} \int_0^t (x-\tau)^{\alpha-1} f_1(x, y, z, \tau) d\tau \\ y(t) - y_0 &= \frac{1-\alpha}{AB(\alpha)} f_2(x, y, z, \tau) + \frac{\alpha}{AB(\alpha)\Gamma(\alpha)} \int_0^t (x-\tau)^{\alpha-1} f_2(x, y, z, \tau) d\tau \\ z(t) - z_0 &= \frac{1-\alpha}{AB(\alpha)} f_2(x, y, z, \tau) + \frac{\alpha}{AB(\alpha)\Gamma(\alpha)} \int_0^t (x-\tau)^{\alpha-1} f_2(x, y, z, \tau) d\tau \end{aligned}\right\} \tag{10.11}$$

By applying the fundamental theorem of fractional calculus, the earlier equations can be converted to fractional integral equation

$$\begin{aligned} x(t) - x(0) = {} & \frac{1-\alpha}{AB(\alpha)} f_1(x(t), y(t), z(t), (t)) \\ & + \frac{\alpha}{AB(\alpha)\Gamma(\alpha)} \int_0^t (t-\tau)^{\alpha-1} f_1(x(\tau), y(\tau), z(\tau), (\tau)) d\tau \end{aligned}$$

$$\begin{aligned} x(t_{n+1}) - x(0) = {} & \frac{1-\alpha}{AB(\alpha)} f_1(x, y, z, t_n) \\ & + \frac{\alpha}{AB(\alpha)\Gamma(\alpha)} \int_0^{t_{n+1}} (t_{n+1}-\tau)^{\alpha-1} f_1(x(\tau), y(\tau), z(\tau), (\tau)) d\tau \\ x(t_{n+1}) - x(0) = {} & \frac{1-\alpha}{AB(\alpha)} f_1(x(\tau), y(\tau), z(\tau), (\tau)) \\ & + \frac{\alpha}{AB(\alpha)\Gamma(\alpha)} \sum_{k=0}^{n} \int_{t_k}^{t_{k+1}} (t_{n+1}-\tau)^{\alpha-1} f_1(x(\tau), y(\tau), z(\tau), (\tau)) \end{aligned} \tag{10.12}$$

Within the interval $[t_k, t_{k+1}]$, the function $f_1(x(\tau), y(\tau), z(\tau), (\tau))$, using the two Lagrange's polynomial, can be approximated as follows:

$$\begin{aligned} P_k(\tau) = {} & \frac{\tau - t_{k-1}}{t_k - t_{k-1}} f_1(x(t_k), y(t_k), z(t_k), \tau(t_k)) \\ & - \frac{\tau - t_k}{t_k - t_{k-1}} f_1(x(t_{k-1}), y(t_{k-1}), z(t_{k-1}), \tau(t_{k-1})) \\ = {} & \frac{(\tau - t_{k-1})}{h} f_1(x(t_k), y(t_k), z(t_k), \tau(t_k)) \\ & - \frac{\tau - t_k}{h} f_1(x(t_k), y(t_k), z(t_k), \tau(t_k)) \end{aligned} \tag{10.13}$$

The earlier approximation can therefore be included in Eq. (10.12) to produce

$$x_{n+1} = x_0 + \frac{1-\alpha}{AB(\alpha)} f_1(x(t_n), y(t_n), z(t_n), (t_n)) + \frac{\alpha}{AB(\alpha)\Gamma(\alpha)} \sum_{k=0}^{n} \times \left(\begin{array}{l} \frac{f_1(x(t_k), y(t_k), z(t_k), (t_k))}{h} \int_{t_k}^{t_{k+1}} (\tau - t_{k-1})(t_{n+1} - \tau)^{\alpha-1} d\tau \\ - \left(\frac{f_1(x(t_{k-1}), y(t_{k-1}), z(t_{k-1}), (t_{k-1}))}{h} \right) \int_{t_k}^{t_{k+1}} (\tau - t_k)(t_{n+1} - \tau)^{\alpha-1} d\tau \end{array} \right) \tag{10.14}$$

Now we let

$$A_{\alpha,k,1} = \int_{t_k}^{t_{k+1}} (\tau - t_{k-1})(t_{n+1} - \tau)^{\alpha-1} d\tau \tag{10.15}$$

$$\text{and } A_{\alpha,k,2} = \int_{t_k}^{t_{k+1}} (\tau - t_k)(t_{n+1} - \tau)^{\alpha-1} d\tau \tag{10.16}$$

These Eqs. (10.15) and (10.16) can be written in this form

$$A_{\alpha,k,1} = h^{\alpha+1} \frac{(n+1-k)^{\alpha}(n-k+2+\alpha) - (n-k)^{\alpha}(n-k+2+2\alpha)}{\alpha(\alpha+1)} \tag{10.17}$$

$$A_{\alpha,k,2} = h^{\alpha+1} \frac{(n+1-k)^{\alpha+1} - (n-k)^{\alpha}(n-k+2+2\alpha)}{\alpha(\alpha+1)} \tag{10.18}$$

Thus, integrating Eqs. (10.17) and (10.18) and replacing them in Eq. (10.12), we obtain

$$\begin{aligned} x_{n+1} = \; & x_0 + \frac{1-\alpha}{AB(\alpha)} f_1(x, y, z, t_n) + \frac{\alpha}{AB(\alpha)\Gamma(\alpha)} \sum_{k=0}^{n} \left(\frac{h^{\alpha} f_1(x, y, z, t_k)}{\Gamma(\alpha+2)} \right. \\ & \left. \times \left((n+1-k)^{\alpha+1}(n-k+2+\alpha) - (n-k)^{\alpha}(n-k+2+2\alpha) \right) \right) \\ & - \left(\frac{h^{\alpha} f_1(x, y, z, t_{k-1})}{\Gamma(\alpha+2)} \left((n+1-k)^{\alpha+1} - (n-k)^{\alpha}(n-k+2+2\alpha) \right) \right) \end{aligned} \tag{10.19}$$

Similarly, we can apply the Atangana–Toufik method for another two equations, we obtain

$$\begin{aligned} y_{n+1} = y_0 + & \frac{1-\alpha}{AB(\alpha)} f_1(x, y, z, t_n) + \frac{\alpha}{AB(\alpha)\Gamma(\alpha)} \sum_{k=0}^{n} \left(\frac{h^{\alpha} f_2(x, y, z, t_k)}{\Gamma(\alpha+2)} \right. \\ & \times \left((n+1-k)^{\alpha+1}(n-k+2+\alpha) \left((n+1-k)^{\alpha+1}(n-k+2+\alpha) \right. \right. \\ & \left. \left. \left. - \; (n-k)^{\alpha}(n-k+2+2\alpha) \right) \right) \right) - \left(\frac{h^{\alpha} f_2(x, y, z, t_{k-1})}{\Gamma(\alpha+2)} \left((n+1-k)^{\alpha+1} \right. \right. \\ & \left. \left. - \; (n-k)^{\alpha}(n-k+2+2\alpha) \right) \right) \end{aligned} \tag{10.20}$$

$$\begin{aligned}z_{n+1} = {} & z_0 + \frac{1-\alpha}{AB(\alpha)} f_1(x,y,z,t_n) + \frac{\alpha}{AB(\alpha)\Gamma(\alpha)} \sum_{k=0}^{n} \left(\frac{h^{\alpha} f_3(x,y,z,t_k)}{\Gamma(\alpha+2)} \right. \\ & \left. \times \left((n+1-k)^{\alpha+1}(n-k+2+\alpha) - (n-k)^{\alpha}(n-k+2+2\alpha) \right) \right) \\ & - \left(\frac{h^{\alpha} f_3(x,y,z,t_{k-1})}{\Gamma(\alpha+2)} \left((n+1-k)^{\alpha+1} - (n-k)^{\alpha}(n-k+2+2\alpha) \right) \right)\end{aligned} \tag{10.21}$$

10.4 Numerical Scheme with Caputo Fractional Derivative

To apply Atangana–Toufik method, we convert to

$$\begin{cases} {}_0^C D_t^{\alpha} x(t) = \frac{1}{3} \times (-(a+1)x(t) + a - c + z(t), y(t)) + ((1-a).(x(t)^2 - y(t)^2)) \\ \qquad + (2.(a+c-z(t)) \times x(t) \times y(t)). \dfrac{1}{3 \times \sqrt{x(t)^2 + y(t)^2}} \\ {}_0^C D_t^{\alpha} y(t) = \frac{1}{3} \times ((c-a-z(t)) \times x(t) - (a+1)y(t) + (2 \times (a-1) \times x(t) \\ \qquad \times y(t) + (a+c-z(t)) \times (x(t)^2 - y(t)^2)) \times \dfrac{1}{3 \times \sqrt{x(t)^2 + y(t)^2}} \\ {}_0^C D_t^{\alpha} z(t) \doteq \frac{1}{2} \times (3 \times x(t)^2 \times y(t) - y(t)^3) - bz(t) \end{cases}$$

To solve the earlier, we transform to

$$\begin{cases} x(t) - x_0 = \int_0^t f_1(x,y,z,\tau)d\tau \\ y(t) - y_0 = \int_0^t f_2(x,y,z,\tau)d\tau \\ z(t) - z_0 = \int_0^t f_3(x,y,z,\tau)d\tau \\ w(t) - w_0 = \int_0^t f_4(x,y,z,\tau)d\tau \end{cases} \tag{10.22}$$

With Caputo fractional derivative, we have The earlier can be converted to

$$\begin{aligned} x(t) - x_0 &= \frac{1}{\Gamma(\alpha)} \int_0^t f_1(x,y,z,\tau)(x-\tau)^{\alpha-1} d\tau \\ y(t) - y_0 &= \frac{1}{\Gamma(\alpha)} \int_0^t f_2(x,y,z,\tau)(x-\tau)^{\alpha-1} d\tau \\ z(t) - z_0 &= \frac{1}{\Gamma(\alpha)} \int_0^t f_3(x,y,z,\tau)(x-\tau)^{\alpha-1} d\tau \\ w(t) - w_0 &= \frac{1}{\Gamma(\alpha)} \int_0^t f_4(x,y,z,\tau)(x-\tau)^{\alpha-1} d\tau \end{aligned}$$

By applying the fundamental theorem of fractional calculus, the earlier equations can be converted to fractional integral equation $x(t) - x(0) = \frac{1}{\Gamma(\alpha)} \int_0^t (t-\tau)^{\alpha-1} f_1(x, y, z, \tau) d\tau$. At the point $t_{n+1, n = 0,1,2}$. So, the earlier equation is of the form as

$$x(t_{n+1}) - x(0) = \frac{1}{\Gamma(\alpha)} \int_0^{t_{n+1}} (t_{n+1} - \tau)^{\alpha-1} f_1(x(\tau), y(\tau), z(\tau), (\tau)) d\tau$$

$$x(t_{n+1}) - x(0) = \frac{1}{\Gamma(\alpha)} \sum_{k=0}^{n} \int_{t_k}^{t_{k+1}} (t_{n+1} - \tau)^{\alpha-1} f_1(x(\tau), y(\tau), z(\tau), (\tau)) d\tau \tag{10.23}$$

Within the interval $[t_k, t_{k+1}]$, the function $f_1(x(\tau), y(\tau), z(\tau), (\tau))$, using the two Lagrange's polynomial, can be approximated as follows:

$$\begin{aligned} P_k(\tau) &= \frac{\tau - t_{k-1}}{t_k - t_{k-1}} f_1(x(t_k), y(t_k), z(t_k), \tau(t_k)) \\ &\quad - \frac{\tau - t_k}{t_k - t_{k-1}} f_1(x(t_{k-1}), y(t_{k-1}), z(t_{k-1}), \tau(t_{k-1})) \\ &= \frac{(\tau - t_{k-1})}{h} f_1(x(t_k), y(t_k), z(t_k), \tau(t_k)) \\ &\quad - \frac{\tau - t_k}{h} f_1(x(t_k), y(t_k), z(t_k), \tau(t_k)) \end{aligned} \tag{10.24}$$

The earlier approximation can therefore be included in Eq. (10.23) to produce

$$x_{n+1} = x_0 + \frac{1}{\Gamma(\alpha)} \sum_{k=0}^{n} \tag{10.25}$$

$$\begin{pmatrix} \frac{f_1(x(t_k), y(t_k), z(t_k), (t_k))}{h} \int_{t_k}^{t_{k+1}} (\tau - t_{k-1})(t_{n+1} - \tau)^{\alpha-1} d\tau \\ - \left(\frac{f_1(x(t_{k-1}), y(t_{k-1}), z(t_{k-1}), (t_{k-1}))}{h} \right) \int_{t_k}^{t_{k+1}} (\tau - t_k)(t_{n+1} - \tau)^{\alpha-1} d\tau \end{pmatrix}$$

Now we let

$$A_{\alpha,k,1} = \int_{t_k}^{t_{k+1}} (\tau - t_{k-1})(t_{n+1} - \tau)^{\alpha-1} d\tau \tag{10.26}$$

and

$$A_{\alpha,k,2} = \int_{t_k}^{t_{k+1}} (\tau - t_k)(t_{n+1} - \tau)^{\alpha-1} d\tau \tag{10.27}$$

These Eqs. (10.26) and (10.27) can be written in this form

$$A_{\alpha,k,1} = h^{\alpha+1} \frac{(n+1-k)^{\alpha}(n-k+2+\alpha) - (n-k)^{\alpha}(n-k+2+2\alpha)}{\alpha(\alpha+1)} \tag{10.28}$$

$$A_{\alpha,k,2} = h^{\alpha+1} \frac{(n+1-k)^{\alpha+1} - (n-k)^{\alpha}(n-k+2+2\alpha)}{\alpha(\alpha+1)} \tag{10.29}$$

Thus, integrating Eqs. (10.28) and (10.29) and replacing them in Eq. (10.24), we obtain

$$\begin{aligned} x_{n+1} = {} & x_0 + \alpha \sum_{k=0}^{n} \left(\frac{h^\alpha f_1(x(t_k), y(t_k), z(t_k), (t_k))}{\Gamma(\alpha+2)} ((n+1-k)^{\alpha+1} \right. \\ & \left. \times \ (n-k+2+\alpha) - (n-k)^\alpha (n-k+2+2\alpha)) \right) \\ & - \left(\frac{h^\alpha f_1(x(t_{k-1}), y(t_{k-1}), z(t_{k-1}), (t_{k-1}))}{\Gamma(\alpha+2)} ((n+1-k)^{\alpha+1} \right. \\ & \left. - \ (n-k)^\alpha (n-k+2+2\alpha)) \right) \end{aligned}$$

Similarly, we can apply the Atangana–Toufik method for another three equations

$$\begin{aligned} y_{n+1} = {} & y_0 + \alpha \sum_{k=0}^{n} \left(\frac{h^\alpha f_2(x(t_k), y(t_k), z(t_k), (t_k))}{\Gamma(\alpha+2)} ((n+1-k)^{\alpha+1} \right. \\ & \left. \times \ (n-k+2+\alpha) - (n-k)^\alpha (n-k+2+2\alpha)) \right) \\ & - \left(\frac{h^\alpha f_2(x(t_{k-1}), y(t_{k-1}), z(t_{k-1}), (t_{k-1}))}{\Gamma(\alpha+2)} ((n+1-k)^{\alpha+1} \right. \\ & \left. - (n-k)^\alpha (n-k+2+2\alpha)) \right) \\ z_{n+1} = {} & z_0 + \alpha \sum_{k=0}^{n} \left(\frac{h^\alpha f_3(x(t_k), y(t_k), z(t_k), (t_k))}{\Gamma(\alpha+2)} \right. \\ & \left. \times ((n-k+2+\alpha) - (n-k)^\alpha (n-k+2+2\alpha)) \right) \\ & - \left(\frac{h^\alpha f_3(x(t_{k-1}), y(t_{k-1}), z(t_{k-1}), (t_{k-1}))}{\Gamma(\alpha+2)} ((n+1-k)^{\alpha+1} \right. \\ & \left. - \ (n-k)^\alpha (n-k+2+2\alpha)) \right) \\ w_{n+1} = {} & w_0 + \alpha \sum_{k=0}^{n} \left(\frac{h^\alpha f_4(x(t_k), y(t_k), z(t_k), (t_k))}{\Gamma(\alpha+2)} ((n+1-k)^{\alpha+1} \right. \\ & \left. \times \ (n-k+2+\alpha) - (n-k)^\alpha (n-k+2+2\alpha)) \right) \\ & - \left(\frac{h^\alpha f_4(x(t_{k-1}), y(t_{k-1}), z(t_{k-1}), (t_{k-1}))}{\Gamma(\alpha+2)} ((n+1-k)^{\alpha+1} \right. \\ & \left. - \ (n-k)^\alpha (n-k+2+2\alpha)) \right) \end{aligned} \tag{10.30}$$

10.5 Numerical Scheme for Caputo–Fabrizio Fractional Derivative

With Caputo and Fabrizio, we have

$$\begin{cases} {}_0^{CF}D_t^{\alpha}x(t) = f_1(x,y,z,t) \\ {}_0^{CF}D_t^{\alpha}y(t) = f_2(x,y,z,t) \\ {}_0^{CF}D_t^{\alpha}z(t) = f_3(x,y,z,t) \\ {}_0^{CF}D_t^{\alpha}w(t) = f_4(x,y,z,t) \end{cases} \tag{10.31}$$

Then we convert to

$$\begin{aligned} x(t) - x_0 &= \frac{1-\alpha}{M(\alpha)} f_1(x,y,z,\tau) + \frac{\alpha}{M(\alpha)} \int_0^t f_1(x,y,z,\tau)d\tau \\ y(t) - y_0 &= \frac{1-\alpha}{M(\alpha)} f_2(x,y,z,\tau) + \frac{\alpha}{M(\alpha)} \int_0^t f_2(x,y,z,\tau)d\tau \\ z(t) - z_0 &= \frac{1-\alpha}{M(\alpha)} f_3(x,y,z,\tau) + \frac{\alpha}{M(\alpha)} \int_0^t f_3(x,y,z,\tau)d\tau \\ w(t) - w_0 &= \frac{1-\alpha}{M(\alpha)} f_4(x,y,z,\tau) + \frac{\alpha}{M(\alpha)} \int_0^t f_4(x,y,z,\tau)d\tau \end{aligned}$$

By applying the fundamental theorem of fractional calculus, the earlier equations can be converted to fractional integral equation

$$\begin{aligned} x(t) - x(0) = {} & \frac{1-\alpha}{M(\alpha)} f_1\left(x(t), y(t), z(t), (t)\right) \\ & + \frac{\alpha}{M(\alpha)} \int_0^t f_1(x(\tau), y(\tau), z(\tau), w(\tau))d\tau \end{aligned} \tag{10.32}$$

$$\begin{aligned} x(t_{n+1}) - x(0) = {} & \frac{1-\alpha}{M(\alpha)} f_1(x,y,z,t_n) \\ & + \frac{\alpha}{M(\alpha)} \int_0^{t_{n+1}} f_1(x(\tau), y(\tau), z(\tau), w(\tau))d\tau \end{aligned} \tag{10.33}$$

$$\begin{aligned} x(t_{n+1}) - x(0) = {} & \frac{1-\alpha}{M(\alpha)} f_1(x(\tau), y(\tau), z(\tau), (\tau)) \\ & + \frac{\alpha}{M(\alpha)} \sum_{k=0}^{n} \int_{t_k}^{t_{k+1}} f_1(x(\tau), y(\tau), z(\tau), w(\tau))d\tau \end{aligned} \tag{10.34}$$

Within the interval $[t_k, t_{k+1}]$, the function $f_1(x(\tau), y(\tau), z(\tau), (\tau))$, using the two Lagrange's polynomial, can be approximated as follows:

$$
\begin{aligned}
P_k(\tau) = & \frac{\tau - t_{k-1}}{t_k - t_{k-1}} f_1(x(t_k), y(t_k), z(t_k), \tau(t_k)) \\
& - \frac{\tau - t_k}{t_k - t_{k-1}} f_1(x(t_{k-1}), y(t_{k-1}), z(t_{k-1}), \tau(t_{k-1})) \\
= & \frac{(\tau - t_{k-1})}{h} f_1(x(t_k), y(t_k), z(t_k), \tau(t_k)) \\
& - \frac{\tau - t_k}{h} f_1(x(t_k), y(t_k), z(t_k), \tau(t_k))
\end{aligned}
$$

The earlier approximation can therefore be included in Eq. (10.34) to produce

$$
\begin{aligned}
x_{n+1} = & \; x_0 + \frac{1-\alpha}{M(\alpha)} f_1(x(t_n), y(t_n), z(t_n), w(t_n)) \\
& + \frac{\alpha}{M(\alpha)} \sum_{k=0}^{n} \left(\frac{f_1(x(t_k), y(t_k), z(t_k), w(t_k))}{h} \right) \int_{t_k}^{t_{k+1}} (\tau - t_{k-1}) d\tau \\
& - \left(\frac{f_1(x(t_{k-1}), y(t_{k-1}), z(t_{k-1}), w(t_{k-1}))}{h} \right) \int_{t_k}^{t_{k+1}} (\tau - t_k) d\tau)
\end{aligned}
$$

Now, we let

$$
\begin{aligned}
A_{\alpha,k,1} = & \int_{t_k}^{t_{k+1}} (\tau - t_{k-1}) d\tau \qquad (10.35) \\
& \text{let } y = \tau - t_{k-1} \\
A_{\alpha,k,1} = & \int_{t_k - t_{k-1}}^{t_{k+1} - t_{k-1}} y dy \\
= & \left. \frac{y^2}{2} \right|_{t_k - t_{k-1}}^{t_{k+1} - t_{k-1}} \\
= & \frac{(t_{k+1} - t_{k-1})^2}{2} - \frac{(t_k - t_{k-1})^2}{2} \\
= & \frac{(2\Delta t)^2}{2} - \frac{(\Delta t)^2}{2} \\
A_{\alpha,k,1} = & \frac{3(\Delta t)^2}{2}
\end{aligned}
$$

and

$$
A_{\alpha,k,2} = \int_{t_k}^{t_{k+1}} (\tau - t_k) d\tau \qquad (10.36)
$$

This Eq. (10.36) can be written in this form

$$
A_{\alpha,k,2} = \frac{(\Delta t)^2}{2} \qquad (10.37)
$$

Thus, integrating Eqs. (10.36) and (10.37) and replacing them in Eq. (10.34), we obtain

$$x_{n+1} = x_0 + \frac{1-\alpha}{M(\alpha)} f_1(x,y,z,t_n) - \frac{\alpha}{M(\alpha)} \sum_{k=0}^{n} \left(\frac{f_1(x,y,z,t_k)}{\Gamma(\alpha+2)} \right) - \left(\frac{f_1(x,y,z,t_{k-1})}{\Gamma(\alpha+2)} ((k-n-1-2\alpha) \right) \tag{10.38}$$

10.6 Existence and Uniqueness Condition for Atangana–Baleanu Fractional Derivative

In this section, we present the existence and uniqueness condition for Atangana–Baleanu derivative, which is defined by

$$x(t) - x_0 = \frac{1-\alpha}{AB(\alpha)} f_1(x,y,z,\tau) + \frac{\alpha}{AB(\alpha)\Gamma(\alpha)} \int_0^t (x-\tau)^{\alpha-1} f_1(x,y,z,\tau) d\tau \tag{10.39}$$

$$y(t) - y_0 = \frac{1-\alpha}{AB(\alpha)} f_2(x,y,z,\tau) + \frac{\alpha}{AB(\alpha)\Gamma(\alpha)} \int_0^t (x-\tau)^{\alpha-1} f_2(x,y,z,\tau) d\tau \tag{10.40}$$

$$z(t) - z_0 = \frac{1-\alpha}{AB(\alpha)} f_3(x,y,z,\tau) + \frac{\alpha}{AB(\alpha)\Gamma(\alpha)} \int_0^t (x-\tau)^{\alpha-1} f_3(x,y,z,\tau) d\tau \tag{10.41}$$

Without loss of generality, we show that $f_1(x,y,z,\tau)$ is Lipschizt in x to achieve this, we assume that

$$\forall t \in T \, \|x(t)\|_\infty = \sup_{t \in T} |x(t)| < M$$

Now we consider

$$C_{\lambda,\beta} = I_\lambda[t_0 - \lambda, t_0 + \lambda] \times B_\beta[x_0 - \beta, x_0 + \beta]$$

In this proof, we consider the following norm

$$\|g\|_\infty = \sup_{t \in T} |g(t)|$$

$$\|f_1(x_1,y,z,\tau) - f_1(x_2,y,z,\tau)\|_\infty = \left\|-ay^2 + ay^2\right\|_\infty = 0$$

So f_1 is Lipchitz.

Let us consider f_3 to be Lipchitz on z
Let $z_1, z_2 \in B_\beta$, thus

$$\begin{aligned}
\|f_3(x_1, y, z_1, \tau) - f_3(x_2, y, z_2, \tau)\|_\infty &= \left\|-cz_1^2 + az_2^2\right\|_\infty \\
&= |c| \left\|z_1^2 - z_2^2\right\|_\infty \\
&= |c| \left\|(z_1 - z_2)(z_1 + z_2)\right\|_\infty \\
&= |c| \left\|(z_1 - z_2)\right\|_\infty \left\|(z_1 + z_2)\right\|_\infty \\
&\leq 2\,|c|\, M\, |z_1 - z_2| \\
&\leq 2\,|c|\, M\, |z_1 - z_2| \\
\|f_3(x_1, y, z_1, \tau) - f_3(x_2, y, z_2, \tau)\|_\infty &= k_3 \|z_1 - z_2\|_\infty
\end{aligned}$$

Now using the similar approach, we prove that $(f_i)_{i\in\{1,2,3,4\}}$ is Lipchitz. Then we define

$$\Gamma\rho = \begin{cases}
\frac{1-\alpha}{AB(\alpha)} f_1(\rho_1, \rho_2, \rho_3, \rho_4, t) + \frac{\alpha}{AB(\alpha)\Gamma(\alpha)} \int_0^t (t-\tau)^{\alpha-1} f_1(\rho_1, \rho_2, \rho_3, \rho_4, t)d\tau \\
\frac{1-\alpha}{AB(\alpha)} f_2(\rho_1, \rho_2, \rho_3, \rho_4, t) + \frac{\alpha}{AB(\alpha)\Gamma(\alpha)} \int_0^t (t-\tau)^{\alpha-1} f_2(\rho_1, \rho_2, \rho_3, \rho_4, t)d\tau \\
\frac{1-\alpha}{AB(\alpha)} f_3(\rho_1, \rho_2, \rho_3, \rho_4, t) + \frac{\alpha}{AB(\alpha)\Gamma(\alpha)} \int_0^t (t-\tau)^{\alpha-1} f_3(\rho_1, \rho_2, \rho_3, \rho_4, t)d\tau \\
\frac{1-\alpha}{AB(\alpha)} f_4(\rho_1, \rho_2, \rho_3, \rho_4, t) + \frac{\alpha}{AB(\alpha)\Gamma(\alpha)} \int_0^t (t-\tau)^{\alpha-1} f_4(\rho_1, \rho_2, \rho_3, \rho_4, t)d\tau
\end{cases}$$

So we evaluate the following

$$\Gamma\rho - X_0 = \begin{cases}
\Gamma_1\rho_1 - x_0 \\
\Gamma_1\rho_1 - y_0 \\
\Gamma_1\rho_1 - z_0 \\
\Gamma_1\rho_1 - w_0
\end{cases}$$

$$\|\Gamma\rho - X_0\|_\infty = \begin{cases}
\|\Gamma_1\rho_1 - x_0\| \\
\|\Gamma_1\rho_1 - y_0\| \\
\|\Gamma_1\rho_1 - z_0\| \\
\|\Gamma_1\rho_1 - w_0\|
\end{cases}$$

The earlier provides

$$\|\Gamma\rho - X_0\| = \begin{cases}
\frac{1-\alpha}{AB(\alpha)} \|f_1\|_\infty + \frac{\lambda^\alpha \|f_1\|_\infty}{AB(\alpha)\Gamma(\alpha)} \\
\frac{1-\alpha}{AB(\alpha)} \|f_2\|_\infty + \frac{\lambda^\alpha \|f_2\|_\infty}{AB(\alpha)\Gamma(\alpha)} \\
\frac{1-\alpha}{AB(\alpha)} \|f_3\|_\infty + \frac{\lambda^\alpha \|f_3\|_\infty}{AB(\alpha)\Gamma(\alpha)} \\
\frac{1-\alpha}{AB(\alpha)} \|f_4\|_\infty + \frac{\lambda^\alpha \|f_4\|_\infty}{AB(\alpha)\Gamma(\alpha)}
\end{cases}$$

We now show that

$$\forall (X,t) \in C_{\lambda,\beta}$$
$$\|\rho\|_{\infty} < M^{'} < \infty$$

Without loss of generality, we prove that

$$\begin{aligned} \|f_3\|_{\infty} &< M^{'} < \infty \\ \|f_3\|_{\infty} &= \left\|y^4 - cz^2 + c\right\|_{\infty} \\ &\leq \left\|y^4\right\|_{\infty} + c\left\|z^2\right\|_{\infty} + \|c\|_{\infty} \\ &\leq M_y^4 + cM_z^{'2} + c < \infty \end{aligned}$$

Using a similar approach, we show that $(f_i)_{i\in\{1,2,3,4\}}$ are bounded, thus $\|\rho\|_{\infty}$ is bounded with this mind.

$$\|\Gamma\rho - X_0\| = \frac{1-\alpha}{AB(\alpha)}M^{'} + \frac{\lambda^{\alpha}M^{'}}{AB(\alpha)\Gamma(\alpha)}$$

Thus, we show that $\Gamma\rho$ is well posed. $\because$ We also show that Γ is Lipschizt.
Let $\rho, \varphi \in C_{\lambda,\beta}$, then

$$\begin{aligned} \|\Gamma\rho - \Gamma\varphi\|_{\infty} &= \left\|\frac{1-\alpha}{AB(\alpha)}(f(\rho,t) - f(\varphi,t)\right\|_{\infty} \\ &+ \frac{\alpha}{AB(\alpha)\Gamma(\alpha)}\int_0^t \left\|((f(\rho,t) - f(\varphi,t))(t-\rho)^{\alpha-1}\right\|_{\infty} d\tau \\ &\leq \frac{1-\alpha}{AB(\alpha)}k\|\rho-\varphi\|_{\infty} + \frac{\alpha}{AB(\alpha)\Gamma(\alpha)}k\frac{\lambda^{\alpha}}{\alpha}\|\rho-\varphi\|_{\infty} \\ &\leq \left(\frac{1-\alpha}{AB(\alpha)} + \frac{\lambda^{\alpha}}{AB(\alpha)\Gamma(\alpha)}\right)k\|\rho-\varphi\|_{\infty} \end{aligned}$$

Γ is a contradiction if the following inequality holds

$$k < \frac{1}{\dfrac{1-\alpha}{AB(\alpha)} + \dfrac{\lambda^{\alpha}}{AB(\alpha)\Gamma(\alpha)}}$$

This completes the proof.

10.7 Existence and Uniqueness Condition for Caputo Fractional Derivative

Here, we examine the existence and uniqueness condition for the Caputo fractional derivative for this system, which is defined by

$$\begin{cases} x(t) - x_0 = \dfrac{1}{\Gamma(\alpha)} \int_0^t f_1(x,y,z,\tau)(x-\tau)^{\alpha-1}d\tau \\ y(t) - y_0 = \dfrac{1}{\Gamma(\alpha)} \int_0^t f_2(x,y,z,\tau)(x-\tau)^{\alpha-1}d\tau \\ z(t) - z_0 = \dfrac{1}{\Gamma(\alpha)} \int_0^t f_3(x,y,z,\tau)(x-\tau)^{\alpha-1}d\tau \\ w(t) - w_0 = \dfrac{1}{\Gamma(\alpha)} \int_0^t f_4(x,y,z,\tau)(x-\tau)^{\alpha-1}d\tau \end{cases} \tag{10.42}$$

Without loss of generality, we show that $f_1(x,y,z,\tau)$ is Lipchitz in x, and to achieve this, we assume that

$$\forall t \in T \, \|x(t)\|_\infty = \sup_{t\in T} |x(t)| < M$$

Now, we consider

$$C_{\lambda,\beta} = I_\lambda[t_0 - \lambda, t_0 + \lambda] \times B_\beta[x_0 - \beta, x_0 + \beta] \tag{10.43}$$

In this proof, we consider the following norm

$$\|g\|_\infty = \sup_{t\in T} |g(t)| \tag{10.44}$$

$$\|f_1(x_1,y,z,\tau) - f_1(x_2,y,z,\tau)\|_\infty = \left\|-ay^2 + ay^2\right\|_\infty = 0 \tag{10.45}$$

So f_1 is Lipchitz.

Let us consider f_3 to be Lipchitz on z

Let $z_1, z_2 \in B_\beta$, thus

$$\begin{aligned} \|f_3(x_1,y,z_1,\tau) - f_3(x_2,y,z_2,\tau)\|_\infty &= \left\|-cz_1^2 + az_2^2\right\|_\infty \qquad (10.46) \\ &= |c| \left\|z_1^2 - z_2^2\right\|_\infty \\ &= |c| \,\|(z_1 - z_2)(z_1 + z_2)\|_\infty \\ &= |c| \,\|(z_1 - z_2)\|_\infty \|(z_1 + z_2)\|_\infty \\ &\leq 2\,|c|\, M\, |z_1 - z_2| \qquad (10.47) \end{aligned}$$

$$\|f_3(x_1,y,z_1,\tau) - f_3(x_2,y,z_2,\tau)\|_\infty = k_3 \|z_1 - z_2\|_\infty \tag{10.48}$$

Now using a similar approach, we prove that $(f_i)_{i\in\{1,2,3,4\}}$ is Lipchitz.

Now we define

$$\Gamma\rho = \begin{cases} \frac{1}{\Gamma(\alpha)} \int_0^t (t-\tau)^{\alpha-1} f_1(\rho_1,\rho_2,\rho_3,\rho_4,t)\, d\tau \\ \frac{1}{\Gamma(\alpha)} \int_0^t (t-\tau)^{\alpha-1} f_2(\rho_1,\rho_2,\rho_3,\rho_4,t)\, d\tau \\ \frac{1}{\Gamma(\alpha)} \int_0^t (t-\tau)^{\alpha-1} f_3(\rho_1,\rho_2,\rho_3,\rho_4,t)\, d\tau \\ \frac{1}{\Gamma(\alpha)} \int_0^t (t-\tau)^{\alpha-1} f_4(\rho_1,\rho_2,\rho_3,\rho_4,t)\, d\tau \end{cases} \tag{10.49}$$

So we evaluate the following

$$\Gamma\rho - X_0 = \begin{cases} \Gamma_1\rho_1 - x_0 \\ \Gamma_1\rho_1 - y_0 \\ \Gamma_1\rho_1 - z_0 \\ \Gamma_1\rho_1 - w_0 \end{cases} \tag{10.50}$$

$$\|\Gamma\rho - X_0\|_\infty = \begin{cases} \|\Gamma_1\rho_1 - x_0\| \\ \|\Gamma_1\rho_1 - y_0\| \\ \|\Gamma_1\rho_1 - z_0\| \\ \|\Gamma_1\rho_1 - w_0\| \end{cases} \tag{10.51}$$

The earlier provides

$$\|\Gamma\rho - X_0\| = \begin{cases} \frac{\lambda^\alpha \|f_1\|_\infty}{\Gamma(\alpha)} \\ \frac{\lambda^\alpha \|f_2\|_\infty}{\Gamma(\alpha)} \\ \frac{\lambda^\alpha \|f_3\|_\infty}{\Gamma(\alpha)} \\ \frac{\lambda^\alpha \|f_4\|_\infty}{\Gamma(\alpha)} \end{cases} \tag{10.52}$$

We now show that

$$\forall (X,t) \in C_{\lambda,\beta}, \|\rho\|_\infty < M < \infty \tag{10.53}$$

Without loss of generality, we prove that

$$\|f_3\|_\infty < M_1 < \infty \tag{10.54}$$

$$\begin{aligned} \|f_3\|_\infty &= \left\|y^4 - cz^2 + c\right\|_\infty \\ &\le \left\|y^4\right\|_\infty + c\left\|z^2\right\|_\infty + \|c\|_\infty \\ &\le M_y^4 + cM_z^2 + c < \infty \end{aligned} \tag{10.55}$$

Using a similar approach, we show that $(f_i)_{i\in\{1,2,3,4\}}$ are bounded; thus, $\|\rho\|_\infty$ is bounded with this mind

$$\|\Gamma\rho - X_0\| = \frac{\lambda^\alpha M}{\Gamma(\alpha)} \tag{10.56}$$

Hence, we show that $\Gamma\rho$ is well posed . We show that Γ is Lipchitz.

Let $\rho, \varphi \in C_{\lambda,\beta}$, then

$$\begin{aligned}
\|\Gamma\rho - \Gamma\varphi\|_\infty &= \frac{\alpha}{\Gamma(\alpha)} \int_0^t \left\|((f(\rho,t) - f(\varphi,t))(t-\rho)^{\alpha-1}\right\|_\infty d\tau \\
&\le \frac{\lambda^\alpha}{AB(\alpha)\Gamma(\alpha)} \|\rho - \varphi\|_\infty \\
&\le \left(\frac{\lambda^\alpha}{\Gamma(\alpha)}\right) k \|\rho - \varphi\|_\infty
\end{aligned} \tag{10.57}$$

Γ is a contradiction if the following inequality holds.

$$k < \frac{1}{\frac{\lambda^\alpha}{\Gamma(\alpha)}} \tag{10.58}$$

Now, under the condition of well posedness and the contradiction condition in accordance with fixed point theorem condition, we can conclude that the system holds exactness and unique solution.

10.8 Existence and Uniqueness Condition for Caputo–Fabrizio Fractional Derivative

Here, we examine the existence and uniqueness condition for Caputo–Fabrizio fractional derivative

$$\begin{cases}
x(t) - x_0 = \dfrac{1-\alpha}{M(\alpha)} f_1(x,y,z,\tau) + \dfrac{\alpha}{M(\alpha)} \int_0^t f_1(x,y,z,\tau)d\tau \\
y(t) - y_0 = \dfrac{1-\alpha}{M(\alpha)} f_2(x,y,z,\tau) + \dfrac{\alpha}{M(\alpha)} \int_0^t f_2(x,y,z,\tau)d\tau \\
z(t) - z_0 = \dfrac{1-\alpha}{M(\alpha)} f_3(x,y,z,\tau) + \dfrac{\alpha}{M(\alpha)} \int_0^t f_3(x,y,z,\tau)d\tau \\
w(t) - w_0 = \dfrac{1-\alpha}{M(\alpha)} f_4(x,y,z,\tau) + \dfrac{\alpha}{M(\alpha)} \int_0^t f_4(x,y,z,\tau)d\tau
\end{cases} \tag{10.59}$$

Without loss of generality, we show that $f_1(x,y,z,\tau)$ is Lipchitz in x to achieve this, we assume that

$$\forall t \in T \, \|x(t)\|_\infty = \sup_{t\in T} |x(t)| < M \tag{10.60}$$

Now, we consider

$$C_{\lambda,\beta} = I_\lambda[t_0 - \lambda, t_0 + \lambda] \times B_\beta[x_0 - \beta, x_0 + \beta] \tag{10.61}$$

In this proof, we consider the following norm

$$\|g\|_\infty = \sup_{t\in T} |g(t)| \tag{10.62}$$

$$\|f_1(x_1, y, z, \tau) - f_1(x_2, y, z, \tau)\|_\infty = \left\|-ay^2 + ay^2\right\|_\infty = 0 \tag{10.63}$$

So f_1 is Lipchitz.

Let us consider f_3 to be Lipchitz on z

Let $z_1, z_2 \in B_\beta$, thus

$$\begin{aligned} \|f_3(x_1, y, z_1, \tau) - f_3(x_2, y, z_2, \tau)\|_\infty &= \left\|-cz_1^2 + az_2^2\right\|_\infty \\ &= |c| \left\|z_1^2 - z_2^2\right\|_\infty \\ &= |c| \left\|(z_1 - z_2)(z_1 + z_2)\right\|_\infty \\ &= |c| \left\|(z_1 - z_2)\right\|_\infty \left\|(z_1 + z_2)\right\|_\infty \\ &\leq 2 |c| M |z_1 - z_2| \end{aligned} \tag{10.64}$$

$$\|f_3(x_1, y, z_1, \tau) - f_3(x_2, y, z_2, \tau)\|_\infty = k_3 \|z_1 - z_2\|_\infty \tag{10.65}$$

Now using the similar approach, we prove that $(f_i)_{i\in\{1,2,3,4\}}$ is Lipchitz.

We define

$$\Gamma\rho = \begin{cases} \frac{1-\alpha}{M(\alpha)} f_1(\rho_1, \rho_2, \rho_3, \rho_4, t) + \frac{\alpha}{M(\alpha)} \int_0^t f_1(\rho_1, \rho_2, \rho_3, \rho_4, t) d\tau \\ \frac{1-\alpha}{M(\alpha)} f_2(\rho_1, \rho_2, \rho_3, \rho_4, t) + \frac{\alpha}{M(\alpha)} \int_0^t f_2(\rho_1, \rho_2, \rho_3, \rho_4, t) d\tau \\ \frac{1-\alpha}{M(\alpha)} f_3(\rho_1, \rho_2, \rho_3, \rho_4, t) + \frac{\alpha}{M(\alpha)} \int_0^t f_3(\rho_1, \rho_2, \rho_3, \rho_4, t) d\tau \\ \frac{1-\alpha}{M(\alpha)} f_4(\rho_1, \rho_2, \rho_3, \rho_4, t) + \frac{\alpha}{M(\alpha)} \int_0^t f_4(\rho_1, \rho_2, \rho_3, \rho_4, t) d\tau \end{cases} \tag{10.66}$$

So we evaluate the following

$$\Gamma\rho - X_0 = \begin{cases} \Gamma_1\rho_1 - x_0 \\ \Gamma_1\rho_1 - y_0 \\ \Gamma_1\rho_1 - z_0 \\ \Gamma_1\rho_1 - w_0 \end{cases} \tag{10.67}$$

$$\|\Gamma\rho - X_0\|_\infty = \begin{cases} \|\Gamma_1\rho_1 - x_0\| \\ \|\Gamma_1\rho_1 - y_0\| \\ \|\Gamma_1\rho_1 - z_0\| \\ \|\Gamma_1\rho_1 - w_0\| \end{cases} \tag{10.68}$$

The earlier provides

$$\|\Gamma\rho - X_0\| = \begin{cases} \dfrac{1-\alpha}{M(\alpha)}\|f_1\|_\infty + \frac{\|f_1\|_\infty}{M(\alpha)} \\ \dfrac{1-\alpha}{M(\alpha)}\|f_2\|_\infty + \frac{\|f_2\|_\infty}{M(\alpha)} \\ \dfrac{1-\alpha}{M(\alpha)}\|f_3\|_\infty + \frac{\|f_3\|_\infty}{M(\alpha)} \\ \dfrac{1-\alpha}{M(\alpha)}\|f_4\|_\infty + \frac{\|f_4\|_\infty}{M(\alpha)} \end{cases} \tag{10.69}$$

We now show that

$$\begin{aligned} &\forall(X,t) \in C_{\lambda,\beta} \\ &\|\rho\|_\infty < M_1 < \infty \end{aligned} \tag{10.70}$$

Without loss of generality, we prove that

$$\begin{aligned} \|f_3\|_\infty &< M_1 < \infty \\ \|f_3\|_\infty &= \left\|y^4 - cz^2 + c\right\|_\infty \\ &\leq \left\|y^4\right\|_\infty + c\left\|z^2\right\|_\infty + \|c\|_\infty \\ &\leq M_{1y}^{\ 4} + cM_{1z}^{\ 2} + c < \infty \end{aligned} \tag{10.71}$$

Using a similar approach, we show that $(f_i)_{i\in\{1,2,3,4\}}$ are bounded; thus, $\|\rho\|_\infty$ is bounded with this mind.

$$\begin{aligned} \|\Gamma\rho - X_0\| &= \frac{1-\alpha}{M(\alpha)}M_1 + \frac{M_1}{M(\alpha)} \\ &= \frac{2-\alpha}{M(\alpha)}M_1 \end{aligned} \tag{10.72}$$

Thus, we show that $\Gamma\rho$ is well posed. We show that Γ is Lipchitz.

Let $\rho, \varphi \in C_{\lambda,\beta}$, then

$$\begin{aligned} \|\Gamma\rho - \Gamma\varphi\|_\infty = &\left\|\frac{1-\alpha}{M(\alpha)}(f(\rho,t) - f(\varphi,t)\right\|_\infty \\ &+ \frac{\alpha}{M(\alpha)}\int_0^t \|((f(\rho,t) - f(\varphi,t))\|_\infty d\tau \leq \frac{1-\alpha}{M(\alpha)}k\|\rho-\varphi\|_\infty \\ &+ \frac{\alpha}{M(\alpha)}k\|\rho-\varphi\|_\infty \leq \frac{1}{M(\alpha)}k\|\rho-\varphi\|_\infty \\ &\leq \left(\frac{1-\alpha}{AB(\alpha)} + \frac{\lambda^\alpha}{AB(\alpha)\Gamma(\alpha)}\right)k\|\rho-\varphi\|_\infty \end{aligned} \tag{10.73}$$

Γ is a contradiction if the following inequality holds.

$$k < \frac{1}{\frac{1}{M(\alpha)}} \tag{10.74}$$

Now under the condition of well posedness and the contradiction condition in accordance with fixed point theorem condition, we can conclude that the system holds exactness and unique solution.

10.9 Conclusion

Using three different differential and integral operators, we investigated a possible extension of an already modified system of mathematical equation able to capture chaotic behaviour. These three models present different dynamical systems due to the properties of their used respective derivatives, namely the Atangana–Baleanu derivative, the Caputo–Fabrizio derivative and the Riemann–Liouville–Caputo derivative. Due to the high non-linearity of the considered problem, we used a suitable numerical scheme to solve this system of equation numerically. In the future work, the existence and uniqueness of the solutions reported for general two-component differential equations will be extended to multidimensional problems.

References

[1] Hadamard, J. (1892). 'Essai sur l'étude des fonctions données par leur développement de Taylor' (PDF). *Journal de mathématiques pures et appliquées 4e série.* 4 (8): 101–186.

[2] Caputo, M. (1967). 'Linear model of dissipation whose Q is almost frequency independent-II'. *Geophysical Journal International.* 13 (5): 529–539.

[3] Metzler, R.; Klafter, J. (2000). 'The random walk's guide to anomalous diffusion: a fractional dynamics approach'. *Physics Reports.* 339 (1): 1–77.

[4] Wheatcraft, S. W.; Meerschaert, M. M. (2008). 'Fractional conservation of mass' (PDF). *Advances in Water Resources.* 31 (10): 1377–1381.

[5] Benson, D.; Wheatcraft, S.; Meerschaert, M. (2000). 'The fractional-order governing equation of Lévy motion'. *Water Resources Research.* 36 (6): 1413–1423.

[6] Näsholm, S. P.; Holm, S. (2011). 'Linking multiple relaxation, power-law attenuation, and fractional wave equations'. *Journal of the Acoustical Society of America.* 130 (5): 3038–3045.

[7] Kober, H. (1940). 'On fractional integrals and derivatives'. *The Quarterly Journal of Mathematics.* os-11 (1): 193–211.

[8] Alkahtani, B. S. T. (2016). 'Chua's circuit model with Atangana–Baleanu derivative with fractional order'. *Chaos, Solitons & Fractals.* 89: 547–551.

[9] Atangana, A.; Koca, I. (2016). 'Chaos in a simple nonlinear system with Atangana– Baleanu derivatives with fractional order'. *Chaos, Solitons & Fractals.* 89: 447–454.

[10] Tateishi, A. A.; Ribeiro, H. V.; Lenzi, E. K. (2017). 'The role of fractional timederivative operators on anomalous diffusion'. *Frontiers in Physics.* 5: 52.

[11] Atangana, A. (2018). 'Non validity of index law in fractional calculus: A fractional differential operator with Markovian and non-Markovian properties'. *Physica A: Statistical Mechanics and its Applications.* 505: 688–706.

[12] Atangana, A.; Gómez-Aguilar, J. F. (2018). 'Decolonisation of fractional calculus rules: Breaking commutativity and associativity to capture more natural phenomena'. *European Physical Journal Plus.* 133(4): 166.

[13] Atangana, A. (2018). 'Blind in a commutative world: Simple illustrations with functions and chaotic attractors'. *Chaos, Solitons & Fractals.* 114: 347–363.

[14] Toufik, M.; Atangana, A. (2017). 'New numerical approximation of fractional derivative with nonlocal and non-singular kernel: Application to chaotic models'. *The European Physical Journal Plus.* 132 (10): 444.

[15] Matsumoto, T. (December 1984). 'A chaotic attractor from Chua's circuit' (PDF). *IEEE Transactions on Circuits and Systems. IEEE.* CAS-31 (12): 1055–1058.

[16] Mishra, J. (2018). 'Fractional hyper-chaotic model with no equilibrium'. *Chaos, Solitons & Fractals.* 116: 43–53.

[17] Mishra, J. (2018). 'A remark on fractional differential equation involving I-function'. *The European Physical Journal Plus.* 133 (2): 36.

11

A New Numerical Method for a Fractional Model of Non-Linear Zakharov–Kuznetsov Equations via Sumudu Transform

Amit Prakash and Hardish Kaur

National Institute of Technology, Kurukshetra-136119, Haryana, India

CONTENTS

11.1 Introduction

There are various important models in many significant areas like control, diffusion, electromagnetism, signal theory, biology, fluid and in many other engineering sciences that are being modelled using fractional derivatives. Fractional derivatives are essential to describe any mathematical model; however, there are many non-linear models in this universe, whose exact solutions are difficult to find or we can say that it is almost impossible to solve a non-linear fractional model. So, it becomes necessary to solve such fractional models numerically. There are many techniques that help to obtain numerical solutions like variational iteration method (VIM) [1,2], homotopy perturbation method (HPM) [3], homotopy analysis method [4], Laplace transform [5,6], q-homotopy analysis transform method (q-HATM) [7] and many more [8–28].

In this article, we discuss the non-linear time-fractional Zakharov–Kuznetsov equations (FZK) of the form given as

$$\frac{\partial^\beta u}{\partial t^\beta} + a\frac{\partial u^p}{\partial x} + b\frac{\partial^3 u^q}{\partial x^3} + c\frac{\partial^3 u^r}{\partial x \partial y^2} = 0, \tag{11.1}$$

where $u = u(x, y, t)$, β is the parameter characterizing the order of the time-fractional derivative $0 < \beta \leq 1$, where a, b and c are arbitrary constants and p, q and r are non-zero integers that manage the conduct of ion-acoustic waves that are weakly non-linear in a plasma comprising cold ions and hot isothermal electrons in the presence of a uniform magnetic field. This equation was first obtained for expressing weakly non-linear ion-acoustic waves in a strongly magnetized lossless plasma in two dimensions. It has been investigated in past years by many techniques like VIM [29], HPM [30], homotopy perturbation transform method (HPTM) [31], new iterative Sumudu transform method (NISTM) [32], etc. This article aims to put up a reliable computational technique to explore non-linear fractional differential equations because of their utilities in mathematical modelling of real-world problems in a more accurate and systematic manner.

In this chapter, we introduced Adomian decomposition Sumudu transform method (ADSTM), which is a combination of Sumudu transform and Adomian decomposition method to find the numerical solution of FZK equations. The positivity of this method is its potential of compiling two strong techniques for numerical solutions of non-linear fractional partial differential equations. We can say that the proposed numerical approach requires less computing time, and it reduces the work of computation in comparison to the established schemes with high accuracy of numerical results.

11.2 Preliminaries

In this section, basic concepts and properties of Sumudu transform and fractional calculus used in the proposed scheme are discussed.

Definition 11.1 The Sumudu transform [33] is defined over a set of functions $A = \{f(t) | \exists M, t_1, t_2 > 0, |f(t)| < Me^{\frac{|t|}{t_j}}$ if $t \in (-1)^j \times [0, \infty)$ by the following formula

$$S[f(t)] = \int_0^\infty f(ut)e^{-t}\mathrm{d}t, u \in (-t_1, t_2).$$

Definition 11.2 The Sumudu transform of Caputo fractional derivative [33] is defined as $S[D_x^{m\alpha} u(x,t)] = s^{-m\alpha} S[u(x,t)] - \sum_{k=0}^{m-1} s^{(-m\alpha+k)} u^k(0,t), m-1 < m\alpha \leq m$.

Definition 11.3 Real function $f(t), t > 0$ is said to be in the space $C_\alpha, \alpha \in R$ [13], if there exists a real number $p, (p > \alpha)$, such that $f(t) = t^p f_1(t)$, where $f_1(t) \in C[0, \infty)$, and it is said to be in the space $C_\alpha^m iff^{(m)} \in C_\alpha, m \in N \bigcup \{0\}$.

Definition 11.4 The Riemann–Liouville fractional integral [34] of order $\alpha \geq 0$ of a function $f(t) \in C_\beta, \beta \geq -1$ is defined as

$I^{\alpha} f(t) = \frac{1}{\Gamma(\alpha)} \int_0^t \frac{f(\tau)}{(t-\tau)^{1-\alpha}} \mathrm{d}\tau,$

$$I^{0} f(t) = f(t),$$

where Γ is the well-known Gamma function.

Definition 11.5 The Caputo fractional derivative [34] of $f(t), f \in C_{-1}^{m}, m \in \mathrm{N}, m > 0$, is defined as
$D^{\alpha} f(t) = I^{m-\alpha} D^{m} f(t) = \frac{1}{\Gamma(m-\alpha)} \int_0^t (t-x)^{m-\alpha-1} f^{m}(x) \mathrm{d}x$, where $m-1 < \alpha \leq m$. The operator D^{α}has the following basic properties:

$$(1) \quad D^{\alpha} I^{\alpha} f(t) = f(t),$$

$$(2) \quad I^{\alpha} D^{\alpha} f(t) = f(t) - \sum_{k=0}^{m-1} f^{(k)}\left(0^{+}\right) \frac{t}{\Gamma(k+1)}, m > 0.$$

11.3 Adomian Decomposition Sumudu Transform Method

To clarify the solution process of ADSTM, consider a general fractional non-linear non-homogeneous partial differential equation of the form:

$$D_t^{\alpha} u(x,t) = Ru(x,t) + Nu(x,t) + g(x,t), n-1 < \alpha \leq n, \tag{11.2}$$

with the initial condition $u(x,0) = f(x)$, $\frac{\partial u(x,0)}{\partial t} = g(x)$.

where $D_t^{\alpha} u(x,t)$ stands for the Caputo's fractional derivative of the function $u(x,t)$, R represents the linear differential operator, N denotes non-linear differential operator and $g(x,t)$ is the source term. First, taking the Sumudu transform on both sides of Eq. (11.2) and making use of the differentiation property of Sumudu transform, it gives

$$u^{-\alpha} S[u(x,t)] - \sum_{k=0}^{m-1} u^{-(\alpha-k)} u^{k}(x,0) = S[Ru(x,t)] + S[Nu(x,t)] + S[g(x,t)].$$

On simplification

$$S[u(x,t)] = \sum_{k=0}^{m-1} u^{k} u^{k}(x,0) + u^{\alpha}\{S[Ru(x,t)] + S[Nu(x,t)] + S[g(x,t)]\}. \tag{11.3}$$

Now, we represent the solution as an infinite series

$$u(x,t) = \sum_{n=0}^{\infty} u_n(x,t), \tag{11.4}$$

and decompose the non-linear operator as

$$Nu\,(x,t) = \sum_{n=0}^{\infty} A_n(u), \tag{11.5}$$

where $A_n(u)$ are the Adomian polynomials of $u_0, u_1, u_2, \ldots, u_n, \ldots$ that are given by

$$A_n\,(u) = \frac{1}{n!}\frac{d^n}{d\lambda^n}\left[N\left(\sum_{i=0}^{\propto} \lambda^i u_i\right)\right]_{|\lambda=0}, n = 0, 1, 2, \ldots$$

using Eqs. (11.4) and (11.5) into Eq. (11.3), we get

$$S\left[\sum_{n=0}^{\infty} u_n(x,t)\right] = \sum_{k=0}^{m-1} u^k u^k\,(x,0) + u^\alpha \left\{S\left[R\left[\sum_{n=0}^{\infty} u_n(x,t)\right]\right]\right.$$
$$\left. + S\left[\sum_{n=0}^{\infty} A_n\,(u)\right] + S\,[g\,(x,t)]\right\}.$$

Comparing both sides, the earlier equation yields the following iterations

$$S\,[u_0\,(x,t)] = \sum_{k=0}^{m-1} u^k u^k\,(x,0)\,,$$
$$S\,[u_1\,(x,t)] = u^\alpha S\,[Ru_0\,(x,t) + A_0\,(u\,(x,t)) + g(x,t)]\,,$$

repeating this process

$$S\,[u_{n+1}\,(x,t)] = u^\alpha S\,[Ru_n\,(x,t) + A_n\,(u\,(x,t))]\,, n \geq 1.$$

Operating inverse of Sumudu transform on both sides of earlier equations

$$u_0\,(x,t) = S^{-1}\left(\sum_{k=0}^{m-1} u^k u^k\,(x,0)\right),$$
$$u_1\,(x,t) = S^{-1}\,(u^\alpha S\,[Ru_0\,(x,t) + A_0\,(u\,(x,t)) + g(x,t)])\,,$$

continuing in this way,

$$u_{n+1}\,(x,t) = S^{-1}\,(u^\alpha S\,[Ru_n\,(x,t) + A_n\,(u\,(x,t))])\,, n \geq 1.$$

Finally, we get the solution in the form given by

$$u\,(x,t) = \lim_{n\to\infty} u_n\,(x,t) = \lim_{n\to\infty} \sum_{j=0}^{n-1} u_j(x,t).$$

11.4 Error Analysis of the Proposed Technique

Theorem 11.1 *If we can find a constant* $0 < \varepsilon < 1$ *such that* $\|v_{m+1}(x,t)\| \leq \varepsilon \|v_m(x,t)\|$ *for each value of* m*. Moreover, if the truncated series* $\sum_{m=0}^{r} v_m(x,t)$ *is employed as a numerical solution* $v(x,t)$*, then the maximum absolute truncated error is determined as*

$$\left\| v(x,t) - \sum_{m=0}^{r} v_m(x,t) \right\| \leq \frac{\varepsilon^{r+1}}{(1-\varepsilon)} \|v_0(x,t)\|.$$

Proof. We have

$$\begin{aligned}\left\| v(x,t) - \sum_{m=0}^{r} v_m(x,t) \right\| &= \left\| \sum_{m=r+1}^{\infty} v_m(x,t) \right\| \\ &\leq \sum_{m=r+1}^{\infty} \|v_m(x,t)\| \\ &\leq \sum_{m=r+1}^{\infty} \varepsilon^m \|v_0(x,t)\| \\ &\leq (\varepsilon)^{r+1} \left[1 + (\varepsilon)^1 + (\varepsilon)^2 + \ldots\right] \|v_0(x,t)\| \\ &\leq \frac{\varepsilon^{r+1}}{(1-\varepsilon)} \|v_0(x,t)\|.\end{aligned}$$

which proves the theorem.

11.5 Test Examples

In this section, the proposed numerical scheme is implemented on two test examples.

Example 1.1 Consider the following FZK (2,2,2) equation as

$$D_t^{\alpha}\vartheta + (\vartheta^2)_x + \frac{1}{8}\left(\vartheta^2\right)_{xxx} + \frac{1}{8}\left(\vartheta^2\right)_{xyy} = 0, 0 < \alpha \leq 1, \tag{11.6}$$

and given initial condition $\vartheta(x,y,0) = \frac{4}{3}\rho sinh^2(x+y)$, where ρ is an arbitrary constant. The exact solution of Eq. (11.6) as derived in [23] is given as

$$\vartheta(x,y,t) = \frac{4}{3}\rho sinh^2(x+y-\rho t).$$

Employing the Sumudu transform on both sides of Eq. (11.6), we obtain

$$\overline{\vartheta} = \frac{4}{3}\rho \sinh^2(x+y) + s^{\alpha} S\left[-\left(\vartheta^2\right)_x - \frac{1}{8}\left(\vartheta^2\right)_{\mathrm{xxx}} - \frac{1}{8}\left(\vartheta^2\right)_{\mathrm{xyy}}\right].$$

Applying the inverse Sumudu transform on both sides, we get

$$\vartheta = \frac{4}{3}\rho \sinh^2(x+y) + S^{-1}\left(s^{\alpha} S\left[-\left(\left(\sum_{k=0}^{\infty}\lambda^k \vartheta_k\right)^2\right)_x - \frac{1}{8}\left(\left(\sum_{k=0}^{\infty}\lambda^k \vartheta_k\right)^2\right)_{xxx} - \frac{1}{8}\left(\left(\sum_{k=0}^{\infty}\lambda^k \vartheta_k\right)^2\right)_{xyy}\right]\right).$$

Solving

$$\begin{aligned}
\vartheta_0 &= \frac{4}{3}\rho \sinh^2(x+y),\\
\vartheta_1 &= S^{-1}\left[s^{\alpha} S(A_0)\right]\\
&= \frac{16\rho^2 t^{\alpha}}{9\Gamma(\alpha+1)}\left[-14\sinh^3(x+y)\cosh(x+y) - 6\sinh(x+y)\cosh^3(x+y)\right],\\
\vartheta_2 &= S^{-1}\left[s^{\alpha} S(A_1)\right]\\
&= \frac{\rho^3 t^{2\alpha}}{\Gamma(2\alpha+1)}\left[549\sinh^4(x+y)\cosh^2(x+y) + 313\sinh^2(x+y)\right.\\
&\quad \left.\times \cosh^4(x+y) + 275\sinh^6(x+y)\right],
\end{aligned}$$

where

$$\begin{aligned}
A_0 &= -(\vartheta_0^2)_x - \frac{1}{8}(\vartheta_0^2)_{xxx} - \frac{1}{8}(\vartheta_0^2)_{xyy},\\
A_1 &= -2\vartheta_1(\vartheta_0)_x - 2\vartheta_0(\vartheta_1)_x - \frac{1}{4}\vartheta_0(\vartheta_1)_{xxx} - \frac{1}{4}\vartheta_1(\vartheta_0)_{xxx}\\
&\quad - \frac{1}{4}\vartheta_0(\vartheta_1)_{xyy} - \frac{1}{4}\vartheta_1(\vartheta_0)_{xyy},
\end{aligned}$$

Figure 11.1a–c represents the exact and approximate solutions and absolute error when $\alpha = 1, t = 0.5$ and $\rho = 0.001$. It can be seen from Figure 11.1a and b that the solution obtained by ADSTM is nearly identical with the exact solution. Here, third-order approximation $u_3(x, y, t)$ is used to evaluate the numerical solution, and the efficiency of ADSTM can be enhanced by evaluating higher-order approximations. We can observe the behaviour of approximate solution for different values of α from Figure 11.2. In Figure 11.2, approximate solution decreases as t increases.

Example 1.2 Consider the following FZK (2,2,2) equation as

$$D_t^{\alpha}\vartheta + (\vartheta^3)_x + 2(\vartheta^3)_{xxx} + 2(\vartheta^2)_{xyy} = 0, 0 < \alpha \leq 1, \tag{11.7}$$

and given initial condition $\vartheta(x, y, 0) = \frac{3}{2}\rho \sinh(\frac{1}{6}(x+y))$, where ρ is an arbitrary constant. The exact solution of Eq. (11.7) as derived in [23] is given as

$$\vartheta(x, y, t) = \frac{4}{3}\rho \sinh^2(x + y - \rho t).$$

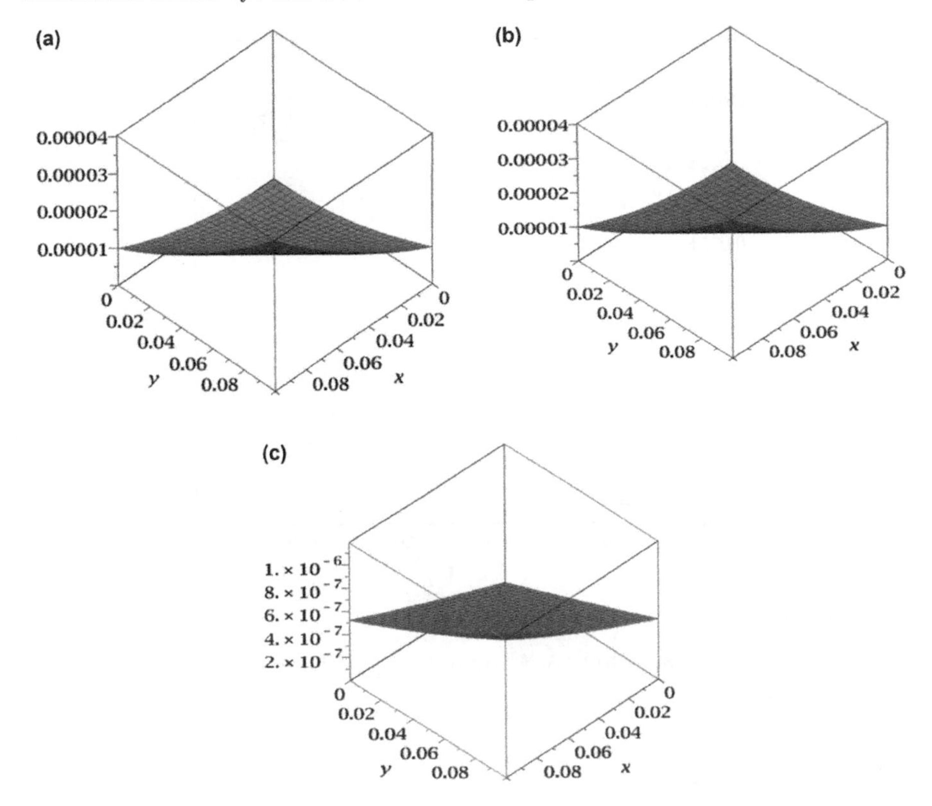

FIGURE 11.1
(a) Exact solution when $\alpha = 1$ for Example 1.1. (b) Approx. solution when $\alpha = 1$ for Example 1.1. (c) Absolute error when $\alpha = 1$ for Example 1.1.

Applying the Sumudu transform on both sides of Eq. (11.7), we obtain

$$\overline{\vartheta} = \frac{3}{2}\rho \sinh\left(\frac{1}{6}(x+y)\right) + s^{\alpha} S\left[-\left(\vartheta^{3}\right)_{x} - 2\left(\vartheta^{3}\right)_{\mathrm{xxx}} - 2\left(\vartheta^{3}\right)_{\mathrm{xyy}}\right].$$

Exerting inverse Sumudu transform on both sides, we have

$$\vartheta = \frac{3}{2}\rho \sinh\left(\frac{1}{6}(x+y)\right) + S^{-1}\left(s^{\alpha} S\left[-\left(\vartheta^{3}\right)_{x} - 2\left(\vartheta^{3}\right)_{\mathrm{xxx}} - 2\left(\vartheta^{3}\right)_{\mathrm{xyy}}\right]\right).$$

Solving, we get

$$\begin{aligned}
\vartheta_0 &= \frac{3}{2}\rho \sinh(\frac{1}{6}(x+y)), \\
\vartheta_1 &= S^{-1}\left[s^{\alpha} S\left(A_0\right)\right] \\
&= \frac{\rho^3 t^{\alpha}}{\Gamma(\alpha+1)}\left[-3\sinh^2\left(\frac{1}{6}(x+y)\right)\cosh\left(\frac{1}{6}(x+y)\right)\right. \\
&\quad \left. -\frac{3}{8}\cosh^3(\frac{1}{6}(x+y))\right],
\end{aligned}$$

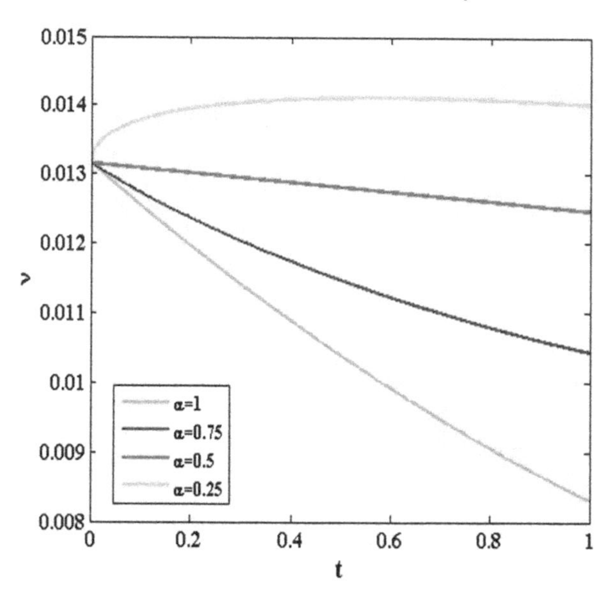

FIGURE 11.2
Plots of approximate solution vs. t when $x = y = 1$ for different values of α for Example 1.1.

$$\begin{aligned}\vartheta_2 &= S^{-1}\left[s^\alpha S\left(A_1\right)\right]\\ &= \frac{\rho^5 t^{2\alpha}}{\Gamma(2\alpha+1)}\left[23\sinh^3\left(\frac{1}{6}(x+y)\right)\cosh^2\left(\frac{1}{6}(x+y)\right)\right.\\ &\quad\left.+\,6\sinh^5\left(\frac{1}{6}(x+y)\right)+0.9\sinh\left(\frac{1}{6}(x+y)\right)\cosh^4\left(\frac{1}{6}(x+y)\right)\right],\end{aligned}$$

where

$$\begin{aligned}A_0 &= -(\vartheta_0^3)_x - 2(\vartheta_0^3)_{xxx} - 2(\vartheta_0^3)_{xyy},\\ A_1 &= -3(\vartheta_0^2)\left(\vartheta_1\right)_x - 12(\vartheta_0^2)\left(\vartheta_1\right)_{xxx} - 6\vartheta_1\vartheta_0\left(\vartheta_0\right)_x - 24\vartheta_0\vartheta_1\left(\vartheta_0\right)_{xxx},\end{aligned}$$

Figure 11.3a–c represents the exact and approximate solutions and absolute error when $\alpha = 1, t = 0.5$ and $\rho = 0.001$. It can be seen from Figure 11.3a and b that the solution obtained by ADSTM is nearly identical with exact solution. Here, third-order approximation $u_3(x, y, t)$ is used to evaluate the numerical solution, and the efficiency of ADSTM can be enhanced by evaluating higher-order approximations. We can observe the behaviour of approximate solution for different values of α from Figure 11.4. In Figure 11.4, approximate solution decreases as t increases.

Table 11.1 shows the comparison of the results obtained from ADSTM with NISTM and HPTM, and a good agreement of results have been found. Tables 11.2 and 11.3 show that the absolute error between the consecutive

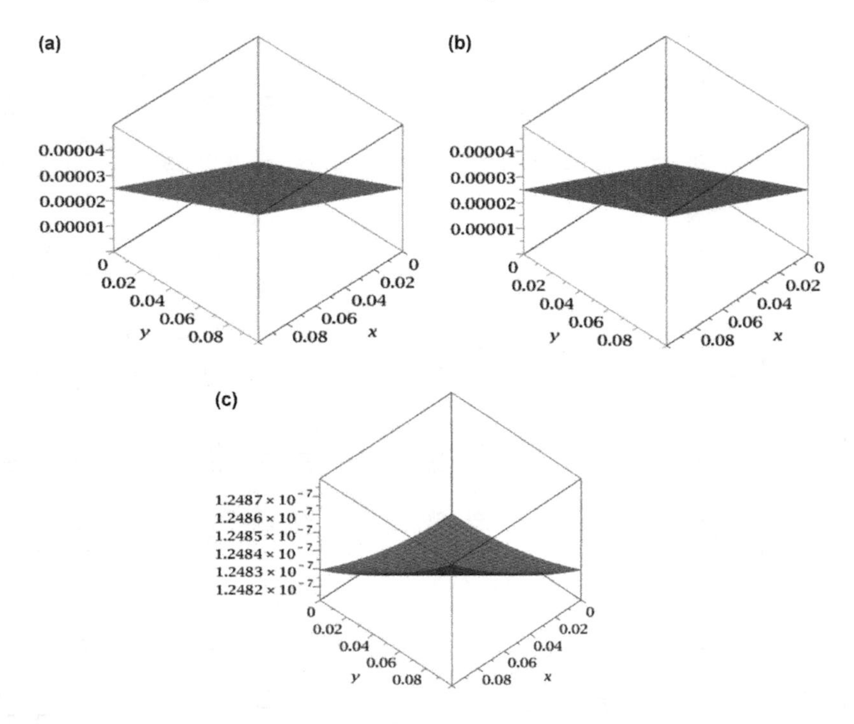

FIGURE 11.3
(a) Exact solution when $\alpha = 1$ for Example 1.2. (b) Approximate solution when $\alpha = 1$ for Example 1.2. (c) Absolute error when $\alpha = 1$ for Example 1.2.

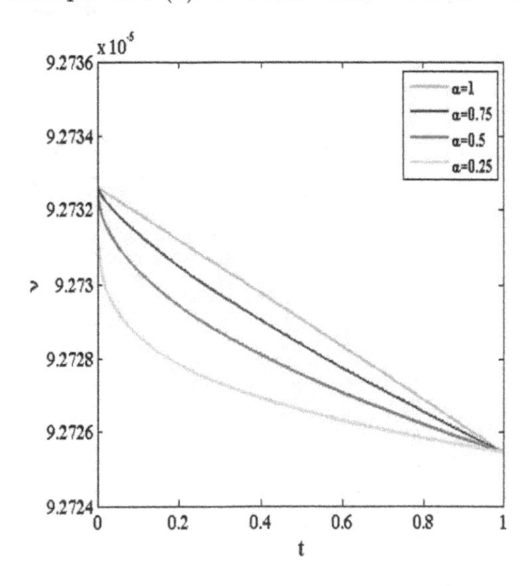

FIGURE 11.4
Plots of approximate solution vs. t when $x = y = 1$ for different values of α for Example 1.2.

TABLE 11.1
Comparison among Absolute Error When $t = 1, \alpha = 1$ **and** $\rho = 0.001$, for Example 1.1

	x \ y	**0**	**0.02**	**0.04**	**0.06**	**0.08**	**0.10**
ADSTM		2.00×10^{-8}	1.4×10^{-7}	3.01×10^{-7}	4.66×10^{-7}	6.35×10^{-7}	8.12×10^{-7}
NISTM[11]	0.00	2.00×10^{-8}	1.4×10^{-7}	3.01×10^{-7}	4.66×10^{-7}	6.35×10^{-7}	8.12×10^{-7}
HPTM [10]		2.00×10^{-8}	1.4×10^{-7}	3.01×10^{-7}	4.65×10^{-7}	6.35×10^{-7}	8.11×10^{-7}
ADSTM		1.40×10^{-7}	3.01×10^{-7}	4.66×10^{-7}	6.35×10^{-7}	8.12×10^{-7}	9.98×10^{-7}
NISTM[11]	0.02	1.40×10^{-7}	3.01×10^{-7}	4.66×10^{-7}	6.35×10^{-7}	8.12×10^{-7}	9.98×10^{-7}
HPTM [10]		1.40×10^{-7}	3.01×10^{-7}	4.65×10^{-7}	6.35×10^{-7}	8.11×10^{-7}	9.97×10^{-7}
ADSTM		3.01×10^{-7}	4.66×10^{-7}	6.35×10^{-7}	8.12×10^{-7}	9.98×10^{-7}	1.2×10^{-6}
NISTM[11]	0.04	3.01×10^{-7}	4.66×10^{-7}	6.35×10^{-7}	8.12×10^{-7}	9.98×10^{-7}	1.2×10^{-6}
HPTM [10]		3.01×10^{-7}	4.66×10^{-7}	6.35×10^{-7}	8.12×10^{-7}	8.11×10^{-7}	9.97×10^{-7}
ADSTM		4.66×10^{-7}	6.35×10^{-7}	8.12×10^{-7}	9.98×10^{-7}	1.2×10^{-6}	1.41×10^{-6}
NISTM[11]	0.06	4.66×10^{-7}	6.35×10^{-7}	8.12×10^{-7}	9.98×10^{-7}	1.2×10^{-6}	1.41×10^{-6}
HPTM [10]		4.66×10^{-7}	6.35×10^{-7}	8.12×10^{-7}	9.98×10^{-7}	1.2×10^{-6}	1.41×10^{-6}
ADSTM		6.35×10^{-7}	8.12×10^{-7}	9.98×10^{-7}	1.2×10^{-6}	1.41×10^{-6}	1.63×10^{-6}
NISTM[11]	0.08	6.35×10^{-7}	8.12×10^{-7}	9.98×10^{-7}	1.2×10^{-6}	1.41×10^{-6}	1.63×10^{-6}
HPTM [10]		6.35×10^{-7}	8.12×10^{-7}	9.98×10^{-7}	1.2×10^{-6}	1.41×10^{-6}	1.63×10^{-6}
ADSTM		8.12×10^{-7}	9.98×10^{-7}	1.2×10^{-6}	1.41×10^{-6}	1.63×10^{-6}	1.88×10^{-6}
NISTM[11]	0.10	8.12×10^{-7}	9.98×10^{-7}	1.2×10^{-6}	1.41×10^{-6}	1.63×10^{-6}	1.88×10^{-6}
HPTM [10]		8.12×10^{-7}	9.98×10^{-7}	1.2×10^{-6}	1.41×10^{-6}	1.63×10^{-6}	1.88×10^{-6}

TABLE 11.2
Difference between Consecutive Iterations, for Example 1.1

x	t	**Exp**.1.1.($\boldsymbol{a} = 1$)		**Exp**.1.1.($\boldsymbol{a} = 0.5$)	
		$\lvert \boldsymbol{u}_2 - \boldsymbol{u}_1 \rvert$	$\lvert \boldsymbol{u}_3 - \boldsymbol{u}_2 \rvert$	$\lvert \boldsymbol{u}_2 - \boldsymbol{u}_1 \rvert$	$\lvert \boldsymbol{u}_3 - \boldsymbol{u}_2 \rvert$
0	0	0	0	0.00000000158	0.000000001000
0.02	0.02	0.000001842391646	0.000001521519261	0.000001873400111	0.000001521267220
0.04	0.04	0.000006906021316	0.000006252966114	0.000006969229611	0.000006251938058
0.06	0.06	0.00001523124531	0.00001422558183	0.00001532906927	0.00001422319267
0.08	0.08	0.00002687970719	0.00002549032028	0.00002701585418	0.00002548587815
0.10	0.10	0.00004193494620	0.00004011916725	0.00004211451789	0.00004011182010

TABLE 11.3
Difference between Consecutive Iterations, for Example 1.2

x	t	**Exp**.1.2.($a = 1$)		**Exp**.1.2.($a = 0.5$)	
		$\lvert u_2 - u_1 \rvert$	$\lvert u_3 - u_2 \rvert$	$\lvert u_2 - u_1 \rvert$	$\lvert u_3 - u_2 \rvert$
0	0	3.75×10^{-11}	3.01×10^{-18}	4.23×10^{-11}	4.20×10^{-18}
0.02	0.02	3.75×10^{-11}	6.01×10^{-18}	4.23×10^{-11}	6.01×10^{-18}
0.04	0.04	3.75×10^{-11}	3.01×10^{-18}	4.23×10^{-11}	1.22×10^{-17}
0.06	0.06	3.76×10^{-11}	9.01×10^{-18}	4.25×10^{-11}	1.81×10^{-17}
0.08	0.08	3.77×10^{-11}	1.37×10^{-17}	4.26×10^{-11}	2.45×10^{-17}
0.10	0.10	3.78×10^{-11}	1.39×10^{-17}	4.28×10^{-11}	3.09×10^{-17}

iterations is negligible and decreases to 0 as the number of iterations increases for distinct values of fractional order α, for examples 1.1 and 1.2. It can be concluded from the results that this method also works for those fractional models that do not have exact solution.

11.6 Conclusion

In this chapter, we implemented a new computational technique ADSTM to find the numerical solution of non-linear FZK equations. The solution for non-linear fractional models is obtained in the form of a convergent series with easily evaluated components. Hence, we conclude that the present scheme is an improvement of standard decomposition method, and the obtained results show that the proposed scheme is reliable and powerful in finding the numerical solutions for wide classes of linear and non-linear fractional partial differential equations without using any restrictive assumptions. Comparisons of results show that the results obtained by ADSTM are in good agreement with NISTM and HPTM.

References

[1] He, J. H. (1997). Variational iteration method for delay differential equations. *Communications in Nonlinear Science and Numerical Simulation*, 2, 235–236.

[2] Prakash, A. and Kumar, M. (2016). He's variational iteration method for the solution of nonlinear Newell-Whitehead-Segel equation. *Journal of Applied Analysis and Computation*, 6(3), 738–748.

[3] Yildinm, A. (2009). An Algorithm for solving the fractional nonlinear Schrödinger equation by means of the homotopy perturbation method. *International Journal of nonlinear Sciences and Numerical Simulation*, 10(4), 445–450.

[4] Liao, S. J. (2004). On the homotopy analysis method for nonlinear problems. *Applied Mathematics and Computation*, 147, 499–513.

[5] Kumar, S., Kocak, H. and Yildirim, A. (2012). A fractional model of gas dynamics equations and its analytical approximate solution using Laplace transform. *Z. Naturforsch*, 67a, 389–396.

[6] Kumar, S., Yildirim, A., Khan, Y. and Wei, L. (2012). A fractional model of the diffusion equation and its analytical solution using Laplace transform. *Scientia Iranica B*, 19(4), 1117–1123.

[7] Prakash, A. and Kaur, H. (2017). Numerical solution for fractional model of Fokker-Planck equation by using q-HATM. *Chaos, Solitons and Fractals*, 105, 99–110.

[8] Gupta, P. K. (2011). Approximate analytical solutions of fractional Benney-Lin equation by reduced differential transform method and the homotopy perturbation method. *Computers and Mathematics with Applications*, 58, 2829–2842.

[9] Sakar, M. G., Uludag, F. and Erdogan, F. (2016). Numerical solution of time-fractional nonlinear PDEs with proportional delays by homotopy perturbation method. *Applied Mathematical Modelling*, 40, 6639–6649.

[10] Prakash, A. (2016). Analytical method for space-fractional telegraph equation by homotopy perturbation transform method. *Nonlinear Engineering-Modelling and Application*, 5(2), 123–128.

[11] Kumar, D., Singh, J. and Baleanu, D. (2017). Analytic study of Allen-Cahn equation of fractional order. *Bulletin of Mathematical Analysis and Applications*, 9(1), 31–40.

[12] Singh, J., Kumar, D. and Nieto Juan, J. (2017). Analysis of an El Nino-Southern Oscillation model with a new fractional derivative. *Chaos, Solitons and Fractals*, 99, 109–115.

[13] Singh, J., Kumar, D., Hammouch, Z. and Atangana, A. (2018). A fractional epidemiological model for computer viruses pertaining to a new fractional derivative. *Applied Mathematics and Computation*, 316, 504–515.

[14] Kumar, S. (2013). Numerical computation of time-fractional Fokker-Planck equation arising in solid state physics and circuit theory. *Zeitschrift fur Naturforschung*, 68a, 1–8.

[15] Prakash, A. and Kumar, M. (2016). Numerical solution of two-dimensional time fractional order biological population model. *Open Physics*, 14, 177–186.

[16] Singh, J., Kumar, D. and Swroop, R. (2016). Numerical solution of time- and space-fractional coupled Burger's equations via homotopy algorithm. *Alexandria Engineering Journal*, 55, 1753–1763.

[17] Prakash, A., Goyal, M. and Gupta, S. (2019). A reliable algorithm for fractional Bloch model arising in magnetic resonance imaging. *Pramana-Journal of Physics*, 92(2), 1–10.

[18] Prakash, A., Goyal, M. and Gupta, S. (2018). Fractional variational iteration method for solving time-fractional Newell-Whitehead-Segel equation. *Nonlinear Engineering-Modelling and Application*, doi:10.1515/nleng-2018-0001.

[19] Prakash, A., Verma, V., Kumar, D. and Singh, J. (2018). Analytic study for fractional coupled Burger's equations via Sumudu transform method. *Nonlinear Engineering-Modelling and Application*, 7(4), 323–332.

[20] Prakash, A. and Kumar, M. (2018). Numerical method for time-fractional gas dynamic equations. *Proceedings of the National Academy of Sciences, India Section A: Physical Sciences*, doi:10.1007/s40010-018-0496-4.

[21] Prakash, A., Kumar, M. and Sharma, K. K. (2015). Numerical method for solving coupled Burgers equation. *Applied Mathematics and Computation*, 260, 314–320.

[22] Prakash, A. and Verma, V. (2019). Numerical method for Fractional model of Newell-Whitehead-Segel equation. *Frontiers of Physics*, 7(15).

[23] Prakash, A. and Kaur, H. (2018). An efficient hybrid computational technique for solving nonlinear local fractional partial differential equations arising in fractal media. *Nonlinear Engineering-Modelling and Application*, 7(3), 229–235, doi:10.1515/nleng-2017-0100.

[24] Prakash, A. and Kaur, H. (2018). q-homotopy analysis transform method for space and time- fractional KdV-Burgers equation. *Nonlinear Science Letters A*, 9(1), 44–61.

[25] Prakash, A. and Kumar, M. (2017). Numerical method for fractional dispersive partial differential equations. *Communications in Numerical Analysis*, 2017(1), 1–18.

[26] Prakash, A., Veeresha, P., Prakasha, D. G. and Goyal, M. (2019). A homotopy technique for a fractional order multi-dimensional telegraph equation via the Laplace transform. *European Physical Journal Plus*, 134, 19, 1–18.

[27] Goyal, M., Prakash, A. and Gupta, S. (2019). Numerical simulation for time-fractional nonlinear coupled dynamical model of romantic and interpersonal relationships. *Pramana-Journal of Physics*, 1–14, doi:10.1007/s12043-019-1746-y.

[28] Kumar, D., Singh, J., Prakash, A. and Swaroop, R. (2019). Numerical simulation for system of time-fractional linear and nonlinear differential equations. *Progress in Fractional Differentiation and Applications*, 5(1), 65–77.

[29] Molliq, R. Y., Noorani, M. S. M., Hashim, I. and Ahmad, R. R. (2009). Approximate solutions of fractional Zakharov-Kuznetsov equations by VIM. *Journal of Computational and Applied Mathematics*, 233, 103–108.

[30] Yildirim, A. and Gulkanat, Y. (2010). Analytical approach to fractional Zakharov-Kuznetsov equations by He's homotopy perturbation method. *Communications in Theoretical Physics*, 53, 1005–1010.

[31] Kumar, D., Singh, J. and Kumar, S. (2014). Numerical computation of nonlinear fractional Zakharov-Kuznetsov equation arising in ion-acoustic waves. *Journal of the Egyptian Mathematical Society*, 22, 373–378.

[32] Prakash, A., Kumar, M. and Baleanu, D. (2018). A new iterative technique for a fractional model of non-linear Zakharov-Kuznetsov equations via Sumudu transform. *Applied Mathematics and Computation*, 334, 30–40.

[33] Watugala, G. K. (1993). Sumudu transform: New integral transform to solve differential equations and control engineering problems. *International Journal of Mathematical Education in Science and Technology*, 24(1), 35–43.

[34] Podlubny, I., *Fractional Differential Equations*, Academic Press, New York, 1999.

12

Chirped Solitons with Fractional Temporal Evolution in Optical Metamaterials

Dépélair Bienvenue, Betchewe Gambo and Justin Mibaille
The University of Maroua

Zakia Hammouch
FST Errachidia, Moulay Ismail University of Meknes

Alphonse Houwe
The University of Maroua Limbe
Nautical Arts and Fisheries Institute

CONTENTS

12.1 Introduction

The propagation of chirped soliton in optical metamaterials has become a wide field of research nowadays [1–21]. The particularity of the optical metamaterials is that, their refractive index and permeability are negative [1]. In the past, the pulse is loosed during the propagation in optical metamaterials because of many kinds of perturbations and properties of materials. Today, with the event of chirped soliton that is able to keep its form after perturbation or collision, it is possible to make pulse uniform in the optical metamaterials. By exciting chirped soliton, it is possible to amplify or reduce a pulse that propagates in optical metamaterials [1–2]. In this work, we consider the non-linear model of Schrödinger, which describes the propagation of wave envelop in optical metamaterials. This model is constituted by a balance between dispersion and non-linearity, including the transverse surface. It shows a non-vanishing pulse that propagates in the system. This kind of pulse is one that can be used in the domain of communication to avoid loss of signal. Some of the non-linear systems such as semiconductor, optical fibre and radar need these types of signals. This model takes into account many terms of non-linearity, which are as follows: cubic non-linearity or Kerr non-linearity, which is widely studied in the literature, quintic non-linearity that brings improvement to Kerr non-linearity, septic non-linearity and nonic non-linearity. The other terms are the group velocity dispersion, the self-steepening and the term that includes the transverse surface. We here apply the variable transformation that conducts two equation systems, namely Bessel equation and non-linear Schrödinger equation. We stress that the fractional-derivative is taken in the Riemann–Liouville sense. Solving the couple of equations, we get the chirping solution that takes into account the different parameters of the model. We also obtain the solution of the model that is the wave envelop mentioned earlier [3–12]. We underline that the obtained chirped solitons are new and give an appreciated stability to signals propagating in many optical metamaterials. On the other hand, since the occurrence of fractional derivatives and integrals, the theories of fractional calculus have undergone a significant and even heated development, which has been primarily contributed by pure but not applied mathematicians; the reader can refer to an encyclopaedic book. In the past few decades, however, applied scientists and engineers realized that differential equations with fractional derivative provided a natural framework for the discussion of various kinds of real problems modelled by the aid of fractional derivative, such as viscoelastic systems, signal processing, diffusion processes, control processing, fractional stochastic systems, allometry in biology and ecology [22–33]. This work is structured as follows: In Section 12.1, an introduction is given; in Section 12.2, we materialize the model and describe the terms of the given equation, and we give three solutions to the different equations derived from the different values of parameters; and in Section 12.3, we will give summary of the precedent task.

12.2 Model Description

We consider the following non-linear Schrödinger equation that describes the propagation of ultra-court waves in dynamic systems

$$i\frac{\partial^{\varepsilon} A}{\partial t^{\varepsilon}}+\lambda A_{xx}+\mu|A|^2A+i\alpha|A|^2A_x+\nu|A|^4A+\delta|A|^6A+\sigma|A|^8A+\frac{1}{2k_0}\nabla_{\perp}^2 A=0, \tag{12.1}$$

where the parameters $\lambda, \mu, \alpha, \nu, \delta, \sigma$ are, respectively, related to the terms of group velocity dispersion, Kerr non-linearity, self-steepening, quintic non-linearity, septic non-linearity and nonic non-linearity; and the last term is considered for transverse directions. The quantity $A(x, y, z, t)$ is the field envelop propagating in the optical metamaterial, and k_0 is its wave number and ε is the order of derivation. Here, we are going to consider that $0 \leq \varepsilon < 1$. Making the following transformation

$$A(x, y, z, t) = f(y, z) \times \Phi(x, t), \tag{12.2}$$

and after separating variables, we get the system of two equations as follows:

$$\nabla_\perp^2 f + 2k_0\gamma f = 0, \tag{12.3}$$

$$\begin{aligned} i\frac{\partial^\varepsilon \Phi}{\partial t^\varepsilon} &+ \lambda\Phi_{xx} + \mu|f.\Phi|^2\Phi + i\alpha|f\Phi|^2\Phi_x + \nu|f\Phi|^4\Phi \\ &+ \delta|f\Phi|^6\Phi + \sigma|f\Phi|^8\Phi - \gamma\Phi = 0. \end{aligned} \tag{12.4}$$

By setting $\gamma = -{q_0}^2$, Eq. (12.3) takes the following form:

$$\nabla_\perp^2 f - 2k_0{q_0}^2 f = 0. \tag{12.5}$$

12.2.1 The Modified Riemann–Liouville Derivative and Bessel's Equation

The Jumarie-modified Riemann–Liouville derivative for a function f is defined by the following

$$f^{(\alpha)}(x) = \frac{1}{\Gamma(-\alpha)}\int_0^x (x-\xi)^{-\alpha-1}(f(x) - f(0))d\xi, \qquad \alpha < 0 \tag{12.6}$$

$$\begin{aligned} f^{(\alpha)}(x) = (f^{(\alpha-1)}(x))' &= \frac{1}{\Gamma(1-\alpha)}\frac{d}{dx} \\ &\int_0^x (x-\xi)^{-\alpha}(f(\xi) - f(0))d\xi, 0 < \alpha < 1 \end{aligned} \tag{12.7}$$

$$f^{(\alpha)}(x) = (f^{(n)}(x))^{(\alpha-n)}, \, n \leq \alpha < n+1, \, n \geq 1. \tag{12.8}$$

Besides, the solutions of Eq. (12.5) are called solutions of Bessel's equation. To get them, let us describe the following procedure:First, we define the Laplacian operator in the d-dimensional fraction surface, which is

$$\nabla_\perp^2 = \frac{\partial^2}{\partial y^2} + \frac{\alpha_1 - 1}{y}\frac{\partial}{\partial y} + \frac{\partial^2}{\partial z^2} + \frac{\alpha_2 - 1}{z}\frac{\partial}{\partial z}, \tag{12.9}$$

with $0 < \alpha_1 < 1$ and $0 < \alpha_2 < 1$.
Moreover, we write $f(y,z) = h(y)g(z)$, which furthermore turns Eq. (12.9) to these two following equations:

$$\left[\frac{\partial^2}{\partial y^2} + \frac{\alpha_1 - 1}{y}\frac{\partial}{\partial y} - 2k_0\gamma_y^2\right] h(y) = 0, \tag{12.10}$$

and

$$\left[\frac{\partial^2}{\partial z^2} + \frac{\alpha_2 - 1}{z}\frac{\partial}{\partial z} - 2k_0\gamma_z^2\right] g(z) = 0. \tag{12.11}$$

where the parameters γ_y and γ_z defined as $\gamma^2 = \gamma_y^2 + \gamma_z^2$ are the wave constants in the y and z directions, respectively.

Now, we are going to derive particular solutions to Eqs. (12.10) and (12.11). Arguing as in [2], we set $h(y) = y^{n_1}u(y)$ and write Eq. (12.10) as

$$\left[y^2\frac{d^2}{dy^2} + y\frac{d}{dy} + (-2k_0\gamma_y^2 - n_1^2)\right] u(y) = 0, \tag{12.12}$$

where $n_1 = 1 - \frac{\alpha_1}{2}$.

Equation (12.12) is of Bessel's type and admits the following solution

$$u(y) = C_1 J_{n1}\left(-i\sqrt{2k_0}\gamma_y y\right) + C_2 Y_{n1}\left(-i\sqrt{2k_0}\gamma_y y\right), \tag{12.13}$$

where J_{n1} and Y_{n1} are Bessel functions of first and second kind of order $n1$, respectively. We then write

$$h(y) = y^{n1}\left[C_1 J_{n1}\left(-i\sqrt{2k_0}\gamma_y y\right) + C_2 Y_{n1}\left(-i\sqrt{2k_0}\gamma_y y\right)\right], n_1 = 1 - \frac{\alpha_1}{2} \tag{12.14}$$

Similarly, the solution to Eq. (12.11) is

$$g(z) = z^{n2}\left[C_3 J_{n2}\left(-i\sqrt{2k_0}\gamma_z z\right) + C_4 Y_{n2}\left(-i\sqrt{2k_0}\gamma_z z\right)\right], n_2 = 1 - \frac{\alpha_2}{2} \tag{12.15}$$

Then, we can write the expression of $f(y,z)$ as follows:

$$\begin{aligned} f(y,z) = y^{n1}z^{n2}&\left[C_1 J_{n1}\left(-i\sqrt{2k_0}\gamma_y y\right) + C_2 Y_{n1}\left(-i\sqrt{2k_0}\gamma_y y\right)\right] \\ &\times \left[C_3 J_{n2}\left(-i\sqrt{2k_0}\gamma_z z\right) + C_4 Y_{n2}\left(-i\sqrt{2k_0}\gamma_z z\right)\right]. \end{aligned} \tag{12.16}$$

The coefficients $C_i, i = 1, ..., 4$ are constants. This partial solution describes how the field behaves in the fractional surface: to have the entire wave expression, we now need to study Eq. (12.4) and have the solution in the propagation direction x.

12.2.2 Solutions of Schrödinger Equation

Now, we have to determine the solutions of Eq. (12.4). For this reason, we set

$$\Phi(x,t) = \rho(\xi).\exp[i(\chi(\xi) - \Omega t)], \tag{12.17}$$

where

$$\xi = x - \frac{ut^{\varepsilon}}{\Gamma(1+\varepsilon)}, \tag{12.18}$$

and $u = 1/v$, with v being the velocity of the wave. Introducing ansatz (12.17) into Eq. (12.4), and then making separation of real and imaginary parts, we get the following system

$$u\rho\chi_{\xi} + \Omega\rho + \lambda\rho_{\xi\xi} - \lambda\rho_{\xi}^2 + \mu|f|^2\rho^3 - \alpha|f|^2\rho^3\chi_{\xi} + \nu|f|^4\rho^5 + \delta|f|^6\rho^7 + \sigma|f|^8\rho^9 - \gamma\rho = 0, \tag{12.19}$$

$$u\rho_{\xi} + 2\lambda\chi_{\xi}\rho_{\xi} + \lambda\rho\chi_{\xi\xi} + \alpha|f|^2\rho^2\rho_{\xi} = 0. \tag{12.20}$$

If we multiply Eq. (12.19) by ρ, then after making integration with respect to ξ, one can obtain the chirp that is given by

$$\delta\omega = -\chi_{\xi} = -\frac{A}{\lambda\rho^2} - \frac{u}{2\lambda} + \frac{\alpha|f|^2\rho^2}{4\lambda}, \tag{12.21}$$

where A is an integration constant.

Introducing χ_{ξ} in (12.19) and then multiplying by ρ and making integration with respect to ξ, one can obtain

$$\rho_{\xi}^2 + a_0\rho^2 + a_1\rho^4 + a_2\rho^6 + a_3\rho^8 + a_4\rho^{10} + \frac{a_5}{\rho^2} + a_6 = 0, \tag{12.22}$$

where the constants $a_i (i = 0, ..., 6)$ are

$$a_0 = \frac{(\Omega - \gamma + \frac{u^2}{4\lambda} - \frac{A\alpha|f|^2}{2\lambda})}{\lambda}, \tag{12.23}$$

$$a_1 = \frac{|f|^2(\mu - \frac{\alpha u}{2\lambda})}{2\lambda}, \tag{12.24}$$

$$a_2 = \frac{|f|^4(\nu + \frac{3\alpha^2}{16\lambda})}{3\lambda}, \tag{12.25}$$

$$a_3 = \frac{\delta|f|^6}{4\lambda}, \tag{12.26}$$

$$a_4 = \frac{\sigma|f|^8}{5\lambda}, \tag{12.27}$$

$$a_5 = \frac{A^2}{\lambda^2}, \tag{12.28}$$

$$a_6 = 2E, \tag{12.29}$$

E is an integration constant.

When we set

$$\rho = \sqrt{\Psi}, \tag{12.30}$$

Eq. (12.22) leads to

$$\Psi_{\xi}^{2}+c_0+c_1\Psi+c_2\Psi^2+c_3\Psi^3+c_4\Psi^4+c_5\Psi^5+c_6\Psi^6=0, \tag{12.31}$$

where the constants $c_i(i=0,...,6)$ are

$$c_0=\frac{4A^2}{\lambda^2}, \tag{12.32}$$

$$c_1=8E, \tag{12.33}$$

$$c_2=\frac{4(\Omega-\gamma+\frac{u^2}{4\lambda}-\frac{A\alpha|f|^2}{2\lambda})}{\lambda}, \tag{12.34}$$

$$c_3=\frac{2|f|^2(\mu-\frac{\alpha u}{2\lambda})}{\lambda}, \tag{12.35}$$

$$c_4=\frac{4|f|^4(\nu+\frac{3\alpha^2}{16\lambda})}{3\lambda}, \tag{12.36}$$

$$c_5=\frac{\delta|f|^6}{\lambda}, \tag{12.37}$$

$$c_6=\frac{4\sigma|f|^8}{5\lambda}. \tag{12.38}$$

12.2.3 Soliton Solution

If $A\neq 0$ (which means $c_0\neq 0$) and $c_i(i=1,...,6)\neq 0$, a solution of Eq. (12.31) is given in the form

$$\Psi(\xi)=\beta+\omega\sqrt{\text{sech}(\eta\xi)}. \tag{12.39}$$

Introducing expression (12.39) into (12.31) (with $c_0\neq 0$), the coefficients of the different power of sech lead to the following parameters

$$\beta=\frac{-c_5}{6c_6};\omega=\omega;\eta=2\omega^2\sqrt{-c_6};c_0=\frac{3c_1c_3^2+80c_1c_4c_6\omega^4}{16(25c_2c_5-6c_3c_4)};c_1=\frac{25c_2c_5-6c_3c_4}{75c_6}. \tag{12.40}$$

Expression (12.39) is the bright or gray soliton (Figures 12.1 and 12.2). The related chirp is given into the next form

$$\delta\omega(\xi)=\frac{-A}{\lambda}\left(\beta+\omega\sqrt{\text{sech}(\eta\xi)}\right)-\frac{u}{2\lambda}+\frac{\alpha|f|^2}{4\lambda}\left(\beta+\omega\sqrt{\text{sech}(\eta\xi)}\right). \tag{12.41}$$

A solution of Eq. (12.4) gives

$$\Phi(x,t)=\sqrt{\beta+\omega\sqrt{\text{sech}\left[\eta\left(x-\frac{ut^{\varepsilon}}{\Gamma(1+\varepsilon)}\right)\right]}}\exp\left[i\left(\chi(x-\frac{ut^{\varepsilon}}{\Gamma(1+\varepsilon)})-\Omega t\right)\right], \tag{12.42}$$

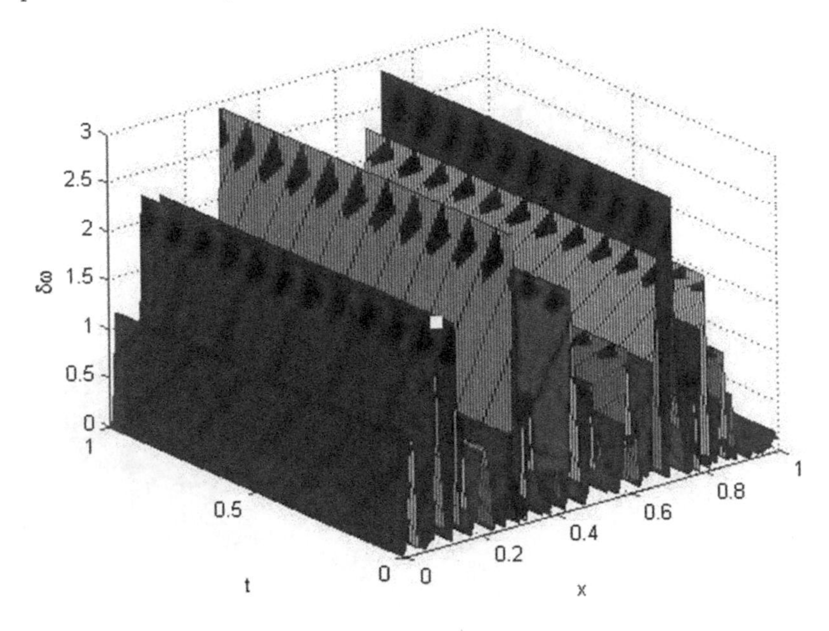

FIGURE 12.1
Graphical representation of the intensity of the field function of Eq. (12.39) which is a bright. The parameters take the following values: $\gamma_y = 0.8$, $\gamma_z = 0.8$, $\alpha_1 = 0.5$, $\alpha_2 = 0.7$, $n_1 = 1 - \alpha_1/2$, $n_2 = 1 - \alpha_2/2$, $C_1 = 1$, $C_2 = 0.5$, $C_3 = 1$, $C_4 = 0.5$, $A = 0.00015$, $E = 0.9$, $u = 1280$, $k_0 = 40$, $\Omega = 3$, $\alpha = 0.8062$, $\lambda = 0.6750$, $\gamma = 0.2236$, $\mu = -2.6885$, $\nu = 0.1174, \delta = 105.3683, \sigma = 108.046$.

and the wave envelop is given by the following expression

$$A(x, y, z, t) = f(y, z) \times \sqrt{\beta + \omega\sqrt{\operatorname{sech}\left[\eta\left(x - \frac{ut^{\varepsilon}}{\Gamma(1+\varepsilon)}\right)\right]}}$$
$$\times \exp\left[i\left(\chi\left(x - \frac{ut^{\varepsilon}}{\Gamma(1+\varepsilon)}\right) - \Omega t\right)\right], \tag{12.43}$$

where

$$f(y, z) = y^{n1} z^{n2} [C_1 J_{n1}(n1, y) + C_2 Y_{n1}(n1, y)][C_3 J_{n2}(n2, z) + C_4 Y_{n2}(n2, z)]. \tag{12.44}$$

In addition, if $E = 0$, $\mu = \dfrac{\alpha u}{2\lambda}$ and $\delta = 0$, which means $c_1 = c_3 = c_5 = 0$, the solution is into the form

$$\Psi(\xi) = \frac{D}{\sqrt{1 + r\cosh(\gamma\xi) + s\sinh(\gamma\xi)}}, \tag{12.45}$$

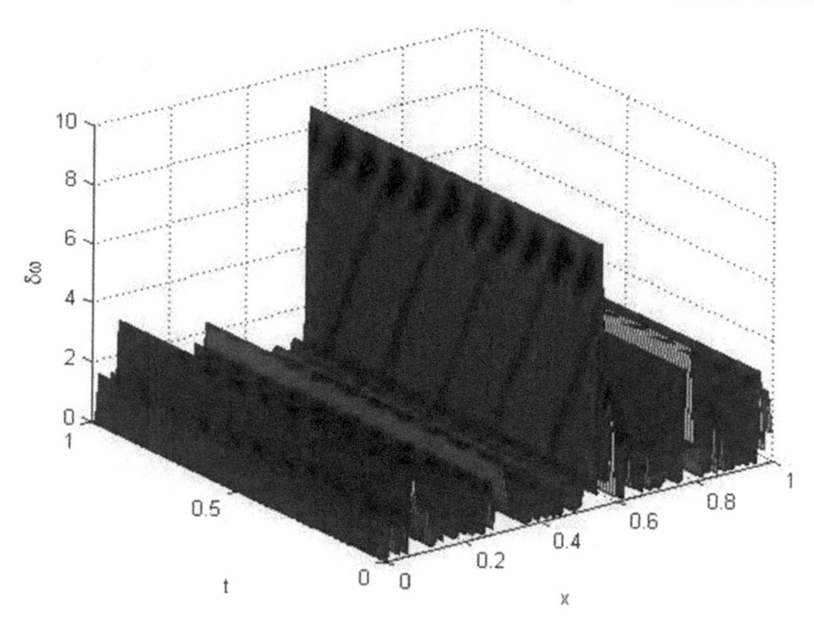

FIGURE 12.2
Graphical representation of the chirp of Eq. (12.41) which is a dark. The parameters take the following values: $\gamma_y = 0.8, \gamma_z = 0.8, \alpha_1 = 0.5, \alpha_2 = 0.7, n_1 = 1 - \alpha_1/2, n_2 = 1 - \alpha_2/2, C_1 = 1, C_2 = 0.5, C_3 = 1, C_4 = 0.5, A = 0.00015, E = 0.9, u = 1280, k_0 = 40, \Omega = -30, \alpha = 0.8062, \lambda = 0.6750, \gamma = 0.2236, \mu = -2.6885, \nu = 0.1174, \delta = 105.3683, \sigma = 108.046$.

with

$$D = \frac{-6c_2}{c_4}; \gamma = \sqrt{-16c_2}; r = \sqrt{1 + s^2 - \frac{36c_2c_6}{5c_4^2}}, \tag{12.46}$$

where $s^2 > \dfrac{36c_2c_6}{5c_4^2} - 1$.

The chirp is given as

$$\delta\omega(\xi) = -\frac{u}{2\lambda} + \frac{\alpha|f|^2}{4\lambda}\Big(\frac{D}{\sqrt{1 + r\cosh(\gamma\xi) + s\sinh(\gamma\xi)}}\Big). \tag{12.47}$$

A solution of Eq. (12.4) gives

$$\Phi(x,t) = \sqrt{\frac{D}{\sqrt{1 + r\cosh(\gamma\xi) + s\sinh(\gamma\xi)}}} \times \exp[i(\chi(x - \frac{ut^{\varepsilon}}{\Gamma(1+\varepsilon)}) - \Omega t)]. \tag{12.48}$$

Then, the wave envelop in Eq. (12.1) is given by the following expression:

$$A(x,y,z,t) = f(y,z)\times\sqrt{\frac{D}{\sqrt{1+r\cosh(\gamma\xi)+s\sinh(\gamma\xi)}}}$$
$$\times\exp[i(\chi(x-\frac{ut^{\varepsilon}}{\Gamma(1+\varepsilon)})-\Omega t)], \quad (12.49)$$

where

$$f(y,z) = y^{n1}z^{n2}[C_1J_{n1}(n1,y)+C_2Y_{n1}(n1,y)][C_3J_{n2}(n2,z)+C_4Y_{n2}(n2,z)]. \quad (12.50)$$

Furthermore, if $r = s$, which means $\frac{36c_2c_6}{5c_4^2} - 1 = 0$ (leading to $c_6 = \frac{5c_4^2}{36c_2}$), we can obtain a kink in the form

$$\Psi(\xi) = \sqrt{\frac{-3c_2}{c_4}[1\pm\tanh(k\xi)]}, \quad (12.51)$$

with $k = 2\sqrt{-c_2}$.

The chirp is given into the form

$$\delta\omega(\xi) = -\frac{u}{2\lambda} + \frac{\alpha|f|^2}{4\lambda}(\sqrt{\frac{-3c_2}{c_4}[1\pm\tanh(k\xi)]}). \quad (12.52)$$

A solution of Eq. (12.4) gives

$$\Phi(x,t) = \sqrt{\frac{-3c_2}{c_4}[1\pm\tanh(k\xi)]}\times\exp[i(\chi(x-\frac{ut^{\varepsilon}}{\Gamma(1+\varepsilon)})-\Omega t)]. \quad (12.53)$$

Then, the wave envelop in Eq. (12.1) is given by the following expression:

$$A(x,y,z,t) = f(y,z)\times\sqrt{\frac{-3c_2}{c_4}[1\pm\tanh(k\xi)]}$$
$$\times\exp[i(\chi(x-\frac{ut^{\varepsilon}}{\Gamma(1+\varepsilon)})-\Omega t)], \quad (12.54)$$

where

$$f(y,z) = y^{n1}z^{n2}[C_1J_{n1}(n1,y)+C_2Y_{n1}(n1,y)][C_3J_{n2}(n2,z)+C_4Y_{n2}(n2,z)]. \quad (12.55)$$

On the other hand, if $\delta = \sigma = 0$ (which means $c_5 = c_6 = 0$) and $c_j(j = 0,...,4) \neq 0$, a solution is in the following form

$$\psi(\xi) = \varsigma + \varrho\,\mathrm{sech}(\theta\xi), \quad (12.56)$$

where the parameters are

$$c_0 = \frac{-c_1c_3}{32c_4}; \quad (12.57)$$

$$\varsigma = \frac{-c_3}{4c_4}; \tag{12.58}$$

$$\theta = \frac{1}{4}\sqrt{-\frac{6c_3^2 - 16c_2c_4}{c_4}}; \tag{12.59}$$

$$\varrho = \frac{\sqrt{2}}{4}\sqrt{\frac{3c_3^2 - 8c_2c_4}{c_4^2}}. \tag{12.60}$$

This last case was treated in [2].

12.3 Conclusion

Through variables separation method, we get three kinds of chirped solitons related to the given model. After making representations, we get notable interesting results. By these results, it is possible to use the obtained chirp to get a pulse that propagates in optical fibre, semiconductors and optical metamaterials. More importantly, these results open way to improve the quality of pulse in the sonar and radar domains.

References

[1] Mehmet E., Mohammad M., Zhou Q., Moshokoa S. P., Biswas A., and Belic M. (2016). Solitons in optical metamaterials with fractional temporal evolution, *Optik* 127, 10879–10897.

[2] Mibaile J., Malwe Boudoue H., Gambo B., and Doka Serge Y. (2018). Optical Chirped Bessel soliton propagating in a metamaterial with a fractal configuration. Preprint submitted to Nolinear Analysis.

[3] Lubin T. (2017). Equations aux dérivées partielles (EDP), Méthode de résolution des EDP par séparation de variables et Applications. 60 cel-01575654.

[4] Jumarie G. (2006). Modified Riemann–Liouville derivative and fractional Taylor series of nondifferentiable functions further results. *Computers and Mathematics with Applications* 51, 1367–1376.

[5] Xu Y. (2015). Soliton propagation through nanoscale waveguides in optical metamaterials. *Optics and Laser Technology* 77, 177–186.

[6] Houria T., Porsezian K., and Grelu P. (2016). Chirped soliton solutions for the generalized nonlinear Schrodinger equation with polynomial nonlinearity and non-Kerr terms of arbitrary order. *Journal of Optics* 18, 075504.

[7] Shihua C., Fabio B., Soto-Crespo Jose M., Yi L., and Philippe G. (2016). Chirped Peregrine solitons in a class of cubic-quintic nonlinear Schrödinger equations. *Physical Review E* 93, 062202.

[8] Houria T., Biswas A., Milovic D., and Belic M. (2016). Chirped femtosecond pulses in the higherorder nonlinear Schrödinger equation with non-Kerr nonlinear terms and cubic-quintic-septic nonlinearities. *Optics Communications.* doi:10.1016/j.optcom.2016.01.005.

[9] Sharma V. K. (2016). Chirped soliton-like solutions of generalized nonlinear Schrödinger equation for pulse propagation in negative index material embedded into a Kerr medium. *Journal of Physics* 90 (11), 1271–1276.

[10] Michelle S., Qin Z., Luminita M., Biswas A., Seithuti P. M., and Milivoj B. (2016). Singular optical solitons in birefringent nano-fibers. *Optik–International Journal for Light and Electron Optics* 127 (20).

[11] Yu-Feng W., Bo T., Ming W., and Hui-Ling Z. (2014). Solitons via an auxiliary function for an inhomogeneous higher-order nonlinear Schrodinger equation in optical fiber communications. *Nonlinear Dynamics* 79 (1), 721–729.

[12] Houria T., Porsezian K., Amitava C., and Tchofo Dinda P. (2016). Chirped solitary pulses for a nonic nonlinear Schrödinger equation on a continuous-wave background. *Physical Review A* 93, 063810.

[13] Alka, Amit G., Rama G., Kumar C. N., and Thokala S. R. (2011). Chirped femtosecond solitons and double-kink solitons in the cubic-quintic nonlinear Schrödinger equation with self-steepening and self-frequency shift. *Physics Review A* 84, 063830.

[14] Krishnan E. V., Muna Al G., Biswas A., Qin Z., Essaid Z., and Milivoj R. B. (2016). Bright and dark optical solitons with kerr and parabolic law nonlinearities by series solution approach. *Journal of Computational and Theoretical Nanoscience* 13, 58–61.

[15] Yanan X., Jose Vega-G., Daniela M., Mirzazadeh M., Mostafa E., Mahmood M. F., Biswas A., and Milivoj B. (2015). Bright and exotic solitons in optical metamaterials by semi-inverse variational principle. *Journal of Nonlinear Optical Physics and Materials* 24 (4), 1550042.

[16] Yanan X., Jun R., and Matthew C. T. (2016). Raman solitons in nanoscale optical waveguides, with metamaterials, having polynomial law nonlinearity. *Emerging Waveguide Technology,* 32–37.

[17] Kibler B. (2007). Propagation non-linèaire d'impulsions ultracourtes dans les fibres optiques de nouvelle génération, thèse de doctorat.tel.archives-ouvertes.fr/tel00169957v2.

[18] Ahmed E., Mohamed Soror A. L., and Marwa M. (2016). Similarity solutions for solving Riesz fractional partial differential equations. *Progress in Fractional Differentiation and Applications* 4, 293–298.

[19] Vasily T. (2006). Electromagnetic fields on fractals. *Modern Physics Letters A* 21 (20), 1587–1600.

[20] Serge D. Y., Justin M., Betchewe, G., and Crepin, K. T. (2017). Optical chirped soliton in metamaterials. *Nonlinear Dynamics* 90 (1), 13–18.

[21] Dumitru B., Alireza K. G., and Ali K. G. (2010). On electromagnetic field in fractional space. *Nonlinear Analysis* 11, 288–292.

[22] Zubair M., Mughal M. J., and Naqvi Q. A. (2010). The wave equation and generalplane wave solutions in fractional space. *Progress in Electromagnetics Research Letters* 19, 137–146, doi:10.2528/PIERL10102103.

[23] Agarwal G.P. (2001). *Nonlinear Fiber Optics*, elsevier.com.

[24] Biswas A., Mirzazadeh M., Mostafa E., Daniela M., and Milivoj B. (2014). Solitons in optical metamaterials by functional variables method and first integral approach. *Frequenz-Berlin* 68 (11–12), 525–530.

[25] El-Sayed A. M. A., El-Mesiry A. E. M., and El-Saka H. A. A. (2007). On the fractional-order logistic equation. *Applied Mathematics Letters* 7 (20), 817–823.

[26] Kaisar R. K., Momina M., and Milivoj R. B. (2015). Nonlinear pulse propagation in optical metamaterials. *Journal of Computational and Theoretical Nanoscience* 11, 4837–4841.

[27] Yusuf P., Ugur K., Yusuf G., and Emine M. (2012). Classification of exact solutions for some nonlinear partial differential equations with generalized evolution. *Abstract and Applied Analysis* 1, 16.

[28] Houria T., Qin Z., Biwas A., Mohammad F. M., Mohammad M., Seithuti M., and Milivoj R. B. (2014). Solitons in optical metamaterials with parabolic law nonlinearityand spatio-temporal dispersion. *Journal of Optoelectronics and Advanced Materials* 16(11–12), 1221–1225.

[29] Bhrawy A. H., Alshaery A. A., Hilal E. M., Khan K. R., Mahmood M. F., and Biswas A. (2014). Optical soliton perturbation with spatio-temporal dispersion in parabolic and dual-power law media by semi-inverse variational principle. *Optik-International Journal for Light and Electron Optics* 125 (17), 4945–4950.

[30] Biwas A., Daniella M., Michelle S., Mohammad F. M., Kaisar R. K., and Russel K. (2012). Optical soliton perturbation in nano fibers with improved nonlinear Schrödinger equation by semi-inverse variational principle. *Journal of Nonlinear Optical Physics & Materials* 21 (4), 1250054.

[31] Younis M. and Rizvi S.T. R. (2015). Optical solitons for ultrashort pulses in nano fibers. *Journal of Nanoelectronics and Optoelectronics* 10 (2), 179–182.

[32] Qin Z. (2014). Analytic study on solitons in the nonlinear fibers with time-modulated parabolic law nonlinearity and Raman effect. *Optik–International Journal for Light and Electron Optics* 13, 3142–3144.

[33] Mibaile J., Malwe B. H., Gambo B., Doka S. Y., and Kofane T. C. (2018). Chirped solitons in derivative nonlinear Schrödinger equation. *Chaos, Solitons and Fractals* 107, 49–54.

13

Controllability on Non-dense Delay Fractional Differential System with Non-Local Conditions

C. Ravichandran
Kongunadu Arts and Science College (Autonomous)

K. Jothimani
Sri Eshwar College of Engineering

Devendra Kumar
University of Rajasthan

CONTENTS

13.1 Introduction

Fractional calculus theory is a range of mathematical study that cultivated from the traditional definition of the calculus derivative and integral operators. It has applications in non-linear oscillations of earthquakes, seepage flow in porous media and in fluid dynamic traffic model. Considerable articles have been concerned to review the presence of smooth solutions of fractional systems [2, 9, 21, 28, 31, 36] and the references therein. To have an efficient interpretation, one can refer the monographs [12, 15, 21, 22, 37]. The authors have discussed the existence and uniqueness result of fractional differential systems with finite delay or infinite delay [3, 13, 18, 23, 26, 29]. Non-local initial conditions, in many cases, are more appropriate and produce better results in applications of physical problems than the classical initial conditions. Some authors have studied fractional differential equations with non-local initial conditions [3, 6, 8, 19, 30, 33, 36].

Controllability views as a vital aspect in the progression of recent mathematical control theory, which leads to very important conclusions regarding finite and infinite dynamical system. From a mathematical point of view, the exact controllability enables to steer the system to arbitrary final state. Wang and zhou [35] have investigated the complete controllability of fractional evolution system, without assuming compactness of the semigroup involved.

It is well known that time delays are frequently encountered in various industrial and practical systems such as chemical processing, bioengineering, fuzzy systems, automatic control, neural networks, circuits and so on. The essence of controllability has already been established for non-densely characterized linear and non-linear systems [1,4–6,8,11,14,16,17,19,20,24,25,27,30, 32,34]. In 2007, Fu et al. [7], addressed the controllability results of non-densely defined fractional neutral system. In 2017, the existence of non-linear evolution equations with fractional order have been discussed by Gu et al. [10] with the aid of non-compact measure. In this connection, in 2018, Ravichandran et al. [24] analyzed the presence of integral solutions of non-linear fractional integro-differential equations. In addition, here we discuss an exact controllability of non-densely defined non-linear fractional integro-differential equations with Hille–Yosida operator.

From the earlier discussions, consider a non-linear fractional integro-differential equation with non-local conditions

$$^{C}D_{0+}{}^{p}z\left(\omega_t\right)=Az(\omega_t)+Bu(\omega_t)+g\Big(\omega_t,z(\omega_t-\rho(\omega_t)),\int_0^{\omega_t}F(\omega_t,s_t,z(s_t-\rho(s_t))ds_t\Big),\omega_t\in J=[0,b]\,, \tag{13.1}$$

$$z(\omega_t)=\varphi(\omega_t)+f(z)(\omega_t),\omega_t\in[-r,0], \tag{13.2}$$

$^{C}D_{0+}{}^{p}$ refers the fractional derivative in Caputo sense with $0<p<1$ and $z(\cdot)$ takes the values in Banach space X and $\|\cdot\|$. A from $D(A)\subset X$ to X, the non-densely linear operator that is closed and g from $[0,b]\times X\times X$ to $D(A)\subset X$, f from $C([-r,0],X)$ to $C([-r,0],X)$ is continuous and non-linear. The term $u(\cdot)$ defines the control of a system, $L^2\left[J,X\right]$, a Banach space of permissible control functions. The operator $B:X\to X$ is linear and bounded. We take $\mathscr{C}:=C([-r,b],X)$.

This chapter illustrates the following: Section 13.2 deals with the basic definitions and fundamental concepts. In Section 13.3, the controllability of (13.1)–(13.2) using Mönch fixed point theorem is discussed.

13.2 Preparatory Results

Here, we discuss some preparatory concepts that are used further.
For $\kappa > 0$, $n = [\kappa]$ (smallest integer $\geq \kappa$) and $z \in (L[0,b], X)$, the fractional order Riemann-Liouville integral is

$$I_{0+}^{\kappa} z(\omega_t) = q_\kappa(\omega_t) * z(\omega_t) = \int_0^{\omega_t} q_\kappa(\omega_t - s_t) z(s_t) ds_t, \qquad \omega_t > 0$$

where $*$ denotes convolution and $q_\kappa(\omega_t) = \frac{\omega_t^{\kappa-1}}{\Gamma(\kappa)}$. If $\kappa = 0$, set $q_0(\omega_t) = \delta(\omega_t)$, the Dirac measure is sticked at the origin.
For $z \in C([0,b], X)$, $\omega_t > 0$, R-L and Caputo fractional derivatives are given by

$$^{R-L}D_{0+}{}^{\kappa} z(\omega_t) = \frac{d^n}{dz^n}\left(q_{n-\kappa}(\omega_t) * z(\omega_t)\right),$$
$$^{C}D_{0+}{}^{\kappa} z(\omega_t) = q_{n-\kappa}(\omega_t) * \frac{d^n}{dz^n} z(\omega_t), \qquad n-1 < \kappa < n.$$

Remark:

1. If $z(\omega_t) \in C^n[0,\infty)$, then
   $^{C}D_{0+}{}^{\kappa} z(\omega_t) = q_{n-\kappa}(\omega_t) * \frac{d^n}{dx^n} z(\omega_t) = I^{n-\kappa} z^n(\omega_t)$,
   $\omega_t > 0, \quad n-1 < \kappa < n$.
2. $^{C}D_{0+}{}^{\kappa}$ (Constant) $= 0$.
3. If z takes values in X, then the integrals that appear in R-L and Caputo fractional derivative are considered as Bochner's sense.

Lemma 13.1 *(see) [10] The Hausdorff non-compact measure $\alpha(\cdot)$ is defined on Ω in X, by $\alpha(\Omega) = \inf\{\in > 0, \Omega$ has finite $\in -$net in $X\}$ which meets*

(1) $\alpha(\Omega_1) \leq \alpha(\Omega_2)$, *for all bounded subsets Ω_1, Ω_2 of X provided $\Omega_1 \subseteq \Omega_2$;*

(2) *for each $y \in X$, $\alpha(\{y\} \cup \Omega) = \alpha(\Omega)$, where $\Omega \subseteq X$ is non-empty;*

(3) $\alpha(\Omega) = 0$ *if and only if Ω is relatively compact in X;*

(4) $\alpha(\Omega_1 + \Omega_2) \leq \alpha(\Omega_1) + \alpha(\Omega_2)$, *where* $\Omega_1 + \Omega_2 = \{y_1 + y_2; y_1 \in \Omega_1, y_2 \in \Omega_2\}$;

(5) $\alpha(\Omega_1 \cup \Omega_2) \leq \max\{\alpha(\Omega_1), \alpha(\Omega_2)\}$;

(6) *for any $\lambda \in R$, $\alpha(\lambda\Omega) \leq |\lambda|\ \alpha(\Omega)$.*

Lemma 13.2 *(see) [10] Let J be the set $[0,b]$, $\{z_n\}_{n=1}^{\infty}$ be a Bochner's defined sequence of integrable functions from J to X with $|z_n(\omega_t)| \leq \widetilde{m}(\omega_t)$, $\omega_t \in J$ and $n \geq 1$, as $\widetilde{m} \in L(J, R^+)$. Also the function $G(\omega_t) = \alpha(\{z_n(\omega_t)\}_{n=1}^{\infty})$ in $L(J, R^+)$ which fulfills*

$$\alpha\left(\left\{\int_0^{\omega_t} z_n(s_t)\, ds_t : n \geq 1\right\}\right) \leq 2\int_0^{\omega_t} G(s_t)\, ds_t.$$

Let $X_0 = \overline{D(A)}$ and A_0 be the representative element of A in X_0 characterized as

$$D(A_0) = \{y \in D(A) : Ay \in X_0\}, \ A_0 y = Ay.$$

Proposition 13.1 *The member A_0 of A develops a Co $-$ semigroup $\{\Re(\omega_t)\}_{\omega_t \geq 0}$ on X_0.*

The following hypothesis introduced for a subsequent analysis.

($H1$) *A from $D(A) \subset X \to X$ fulfills the Hille–Yosida condition, that is there exists two constants $r \in R, \rho$ provided $(r, +\infty) \subseteq \rho(A)$, and for all $\lambda > r$, $n \geq 1$,*

$$\sup(\lambda - r)^n \left\|(\lambda I - A)^{-n}\right\|_{L(X)} \leq \sigma.$$

($H2$) *$\{\Re(\omega_t)\}_{\omega_t \geq 0}$ is uniformly bounded, so there is a $M > 1$ with $\sup_{\omega_t \in [0,+\infty]} |\Re(\omega_t)| < M$. Also for $\omega_t > 0$, $\Re(\omega_t)$ is continuous uniformly.*

For any $z \in X$, $S_p(\omega_t)$ and $T_p(\omega_t), \omega_t \geq 0$ is defined as

$$S_p(\omega_t) = \int_0^{\omega_t} \psi_p(\theta)\, \Re(\omega_t^p \theta)\, d\theta, \qquad T_p(\omega_t) = p\int_0^{\infty} \theta\psi_p(\theta)\, \Re(\omega_t^p \theta)\, d\theta$$

and for $\theta \in (0, \infty)$

$$\psi_p(\theta) = \frac{1}{p}\theta^{(-1-\frac{1}{p})} W_p\left(\theta^{-\frac{1}{p}}\right) \geq 0,$$
$$W_p(\theta) = \frac{1}{\pi}\sum_{n=1}^{\infty}(-1)^{n-1}\theta^{-np-1}\frac{\Gamma(np+1)}{n!}\sin(n\pi p).$$

Here, ψ_p is a probability density function on $(0, \infty)$, that is

$$\psi_p(\theta) \geq 0, \theta \in (0, \infty), \text{ and the total probability is one.}$$

Lemma 13.3 *(see) [24] From Hypothesis* (H_2),

(i) *For any* $\omega_t \geq 0$, $S_p(\omega_t)$ *and* $T_p(\omega_t)$ *are linear and bounded operators, that is for any* $z \in X$,

$$\|S_p(\omega_t) z\| \leq M \|z\| \text{ and } \|T_p(\omega_t) z\| \leq \frac{M_p}{\Gamma(1+p)} \|z\|.$$

(ii) $T_p(\omega_t)$ *is uniformly continuous for* $\omega_t \geq 0$.

(iii) *For fixed* $\omega_t \geq 0$, $\{S_p(\omega_t)\}_{\omega_t>0}$ *and* $\{T_p(\omega_t)\}_{\omega_t>0}$ *are strongly continuous, that is, for* $y \in z_0$, $0 < \omega_{t1} < \omega_{t2} \leq b$,

$$\begin{aligned} &|S_p(\omega_{t1}) y - S_p(\omega_{t2}) y| \to 0 \text{ and} \\ &|T_p(\omega_{t1}) y - T_p(\omega_{t2}) y| \to 0 \text{ as } \omega_{t2} \to \omega_{t1}. \end{aligned} \tag{13.3}$$

Lemma 13.4 *Let* D *be a convex and closed subset of* X*; also* $0 \in D$*. If* F *from* $\overline{D}$ *to* X *is continuous and of Mönch type, that is,* F *fulfills the condition* $u \subseteq \overline{D}$, u *is countable and* $u \subseteq \overline{co}(\{0\} \cup F(u)) \Rightarrow \overline{u}$ *is compact. Then,* F *has at least one fixed point.*

13.3 Results on Controllability

The subsidiary problem of (13.1)–(13.2) is

$$\begin{aligned} {}^C D_{0+}{}^P z(\omega_t) &= A_0 z(\omega_t) + g\Big(\omega_t, z(\omega_t - \rho(\omega_t)), \\ &\int_0^{\omega_t} F(\omega_t, s_t, z(s_t - \rho(s_t)))ds_t\Big), \quad \omega_t \in J \\ z(\omega_t) &= \varphi(\omega_t) + f(z)(\omega_t), \quad \omega_t \in [-r, 0]. \end{aligned} \tag{13.4}$$

We define the integral solution of (13.4) as

$$\begin{aligned} z(\omega_t) = {} & I_{0+}^{1-p} S_p(\omega_t)[\varphi(0) + f(z)(0)] \\ & + \int_0^{\omega_t} S_p(\omega_t - s_t) g\Big(s_t, z(s_t - \rho(s_t)), \int_0^{s_t} F(s_t, \tau_t, z(\tau_t - \rho(\tau_t)))d\tau_t\Big) ds_t. \end{aligned} \tag{13.5}$$

If g assumes values in X_0, then (13.5) becomes

$$z(\omega_t) = I_{0+}^{1-p} S_p(\omega_t)[\varphi(0) + f(z)(0)] + \lim_{\lambda \to +\infty} \int_0^{\omega_t} S_p(\omega_t - s_t) C_\lambda g\Big(s_t, z(s_t - \rho(s_t)), \int_0^{s_t} F(s_t, \tau_t, z(\tau_t - \rho(\tau_t)))d\tau_t\Big) ds_t. \tag{13.6}$$

where $C_\lambda = \lambda(\lambda I - A)^{-1}$. Since $\lim_{\lambda \to +\infty} C_\lambda z = z$, for $z \in X_0$. When g assigns values in X, not in X_0, so the limit in (13.6) exists.

Theorem 13.1 *$z(\omega_t)$ is a solution of (4) iff, for $\omega_t \in J$ and $\varphi(0) + f(z)(0) \in X_0$,*

$$z(\omega_t) = I_{0+}^{1-p} S_p(\omega_t)[\varphi(0) + f(z)(0)] + \lim_{\lambda \to +\infty} \int_0^{\omega_t} S_p(\omega_t - s_t) C_\lambda g\Big(s_t, z(s_t - \rho(s_t)), \int_0^{s_t} F(s_t, \tau_t, z(\tau_t - \rho(\tau_t)))d\tau_t\Big) ds_t.$$

Proof. It is adequate to make sure for $\varphi(0) + f(z)(0) = 0$, because it is true when $g = 0$.

Step I: For a continuous as well as differentiable function g, for $\omega_t \in J$, we have

$$z_\lambda(\omega_t) = \int_0^{\omega_t} S_p(\omega_t - s_t) C_\lambda g\Big(s_t, z(s_t - \rho(s_t)), \int_0^{s_t} F(s_t, \tau_t, z(\tau_t - \rho(\tau_t)))d\tau_t\Big) ds_t$$

$$= \psi_p^0(\omega_t) C_\lambda g\Big(0, z(0 - \rho(0)), \int_0^{s_t} F(0, 0, z(0 - \rho(0)))\, d\tau_t\Big) + \int_0^{\omega_t} \psi_p^0(\omega_t - s_t) C_\lambda g'\Big(s_t, z(s_t - \rho(s_t)), \int_0^{s_t} F(s_t, \tau_t, z(\tau_t - \rho(\tau_t)))d\tau_t\Big) ds_t.$$

By Lemma 13.3, for $\omega_t \in J$,

$$z(\omega_t) = \lim_{\lambda \to +\infty} z_\lambda(\omega_t)$$

$$= \psi_p(\omega_t)\, g\Big(0, z(0-\rho(0)), \int_0^{s_t} F(0,0,z(0-\rho(0)))d\tau_t\Big)$$

$$+ \int_0^{\omega_t} \psi_p(\omega_t - s_t)\, g'\Big(s_t, z(s_t-\rho(s_t)), \int_0^{s_t} F(s_t,\tau_t,z(\tau_t-\rho(\tau_t)))d\tau_t\Big)ds_t$$

$$= A\Big(I_{0+}^p \psi_p(\omega_t)\, g\Big(0, z(0-\rho(0)), \int_0^{s_t} F(0,0,z(0-\rho(0)))d\tau_t\Big)\Big)$$

$$+ \frac{\omega_t^p}{\Gamma(1+p)} g\Big(0, z(0-\rho(0)), \int_0^{s_t} F(0,0,z(0-\rho(0)))d\tau_t\Big)$$

$$+ \int_0^{\omega_t} \Big[A\left(I_{0+}^p \psi_p(\omega_t - s_t)\right) + \frac{(\omega_t - s_t)^p}{\Gamma(1+p)}\Big]$$

$$(\times)\; g'\Big(s_t, z(s_t-\rho(s_t)), \int_0^{s_t} F(s_t,\tau_t,z(\tau_t-\rho(\tau_t)))d\tau_t\Big)ds_t$$

$$= A\left(I_{0+}^p z(\omega_t)\right) + I_{0+}^p g\Big(\omega_t, z(\omega_t-\rho(\omega_t)), \int_0^{s_t} F(s_t,\tau,z(\tau_t-\rho(\tau_t)))d\tau_t\Big),$$

where

$$\psi_p(\omega_t)\, y = \lim_{\lambda\to+\infty} \int_0^{\omega_t} S_p(\omega_t - s_t)\, C_\lambda y ds_t$$

$$= \lim_{\lambda\to+\infty} \int_0^{\omega_t} S_p(s_t)\, C_\lambda y ds_t.$$

Also $\psi_p^0(\omega_t)\, y = \int_0^{\omega_t} S_p(s_t)\, y ds_t$.

Step II: Using continuous differentiable function g_n, approximating g such that n approaches to ∞,

$$\sup_{\omega_t\in J}\Big| g\Big(\omega_t, z(\omega_t-\rho(\omega_t)), \int_0^{s_t} F(s_t,\tau_t,z(\tau_t-\rho(\tau_t)))d\tau_t\Big)$$

$$- g_n\Big(\omega_t, z(\omega_t-\rho(\omega_t)), \int_0^{s_t} F(s_t,\tau_t,z(\tau_t-\rho(\tau_t)))d\tau_t\Big)\Big| \to 0.$$

Letting $z_n(\omega_t) = \lim_{\lambda\to+\infty} \int_0^{\omega_t} S_p(s_t)\, C_\lambda g_n\Big(s_t, z(\tau_t-\rho(\tau_t)),$ $\int_0^{s_t} F(s_t,\tau_t,z(\tau_t-\rho(\tau_t)))d\tau_t\Big)ds_t$, we get

$$z_n(\omega_t) = A\left(I_{0+}^p z_n(\omega_t)\right) + I_{0+}^p g_n\Big(\omega_t, z(\omega_t - \rho(\omega_t)), \int_0^{s_t} F(s_t, \tau_t, z(\tau_t - \rho(\tau_t)))d\tau_t\Big).$$

Then

$$|z_n(\omega_t) - z_m(\omega_t)| \le \frac{M\sigma b^p}{\Gamma(p)} \|g_n - g_m\|,$$

which gives $\{z_n\}$ is a Cauchy sequence. By taking limit

$$z(\omega_t) = A\left(I_{0+}^p z(\omega_t)\right) + I_{0+}^p g\Big(\omega_t, z(\omega_t - \rho(\omega_t)), \int_0^{\omega_t} F(\omega_t, s_t, z(s_t - \rho(s_t)))ds_t\Big),$$

for $\omega_t \in J$.

Using Theorem 13.1, we give a mild solution of the non-linear Cauchy problem (13.1)–(13.2).

Definition 13.1 $z(\omega_t)$ is said to be a solution of (13.1) – (13.2) if

$$z(\omega_t) = I_{0+}^{1-p} S_p(\omega_t)[\varphi(0) + f(z)(0)] + \lim_{\lambda\to+\infty} \int_0^{\omega_t} S_p(\omega_t - s_t) Bu(s_t) ds_t$$

$$+ \lim_{\lambda\to+\infty} \int_0^{\omega_t} S_p(\omega_t - s_t) C_\lambda g\Big(s_t, z(s_t - \rho(s_t)), \int_0^{s_t} F(s_t, \tau_t, z(\tau_t - \rho(\tau_t)))d\tau_t\Big) ds_t.$$

is satisfied.

Further, we introduce the subsequent suppositions.

($H3$) For every $\omega_t \in J$, $y_1, y_2 \in X$, $g(\omega_t, \cdot, \cdot) : X \to X$ is continuous and $g(\cdot, y_1, y_2)$ from J to X is strongly measurable.

($H4$) (i) The function $g(\omega_t, s_t, \cdot) : X \to X$ is almost continuous everywhere $(\omega_t, s_t) \in J$, and for $z \in X$, $g(\cdot, \cdot, z) : J \to X$ is measurable. Also $\eta : J \to R^+$ with $\sup_{\omega_t \in J} \int_0^{\omega_t} \eta(\omega_t, s_t) ds_t = \eta^* < \infty$, provided $\|g(\omega_t, s_t, z)\| \le \eta(\omega_t, s_t) \|z\|$.

(ii) For $D_1 \in X$, and $0 \le s_t \le \omega_t \le b$, $G : J \to R^*$, confirms $\alpha(g(\omega_t, s_t, D_1)) \le G(\omega_t, s_t)\alpha(D_1)$, where $\sup_{\omega_t \in J} \int_0^{\omega_t} G(\omega_t, s_t) ds_t = G^* < \infty$.

$(H5)$ There is a $m_g^* \in L(J, R^+)$, provided $I_{0+}^p m_g^*(\omega_t) \in C(J, R^+)$, $\lim_{\lambda \to +\infty} I_{0+}^p m_g^*(\omega_t) = 0$, with $\sup_{\omega_t \in J} g\Big(\omega_t, z(\omega_t - \rho(\omega_t)), \int_0^{s_t} F(s_t, \tau_t, z(\tau_t - \rho(\omega_t)))d\tau_t\Big) = m_g^*(\omega_t)$, where $||g(\omega_t, y_1, y_2)|| \leq m_g^*(\omega_t)\,\eta\Big[||y_1|| + ||y_2||\Big]$, for all $y_1, y_2 \in X$ and $\omega_t \in J$.

$(H6)$ For some constant $h_g^* > 0$, provided for any bounded $D_1, D_2 \subseteq X$, $\alpha(g(\omega_t, D_1, D_2)) \leq l_g^* \alpha(D_1, D_2)$, almost everywhere $\omega_t \in J$.

$(H7)$ $f : \mathscr{C} \to C([-r, 0], X)$ is a continuous function and there exists a positive constant l_g, N such that $||f(z) - f(x)||_{C([-r,0],X)} \leq l_g||z - x||_{\mathscr{C}}$, for every $z, x \in \mathscr{C}$ and $||f(z)||_{C([-r,0],X)} \leq N$ for every $z \in \mathscr{C}$.

$(H8)$ The operator $W : L^2(J, X) \to X$ is characterized by $W = \int_0^b u(\omega_t, s_t)\,Bu(s_t)\,ds_t$. Also, W^{-1} exists and assumes values in $L^2(J, X)\,|\ker W$ and for $M_2, M_3 \geq 0$, provided $\|B\| \leq M_2$, $\|W^{-1}\| \leq M_3$.

$(H9)$ There is some $l_f^* > 0$, provided for any bounded $z \in X$, $\alpha(u(z, \omega_t)) \leq l_f^* \nu(z, \omega_t)\alpha(z(u))$, almost everywhere $\mu \in J$ with $\sup_{\omega_t \in J} \int_0^{\omega_t} \nu(z, s_t)\,ds_t = \nu^* < \infty$.

Now we define the control term for any $z \in X$ as follows:

$$u(z, \omega_t) = W^{-1}\left[z_1 - T_p(\omega_t)[\varphi(0) + f(z)(0)] - \lim_{\lambda \to +\infty} \int_0^{\omega_t} S_p(\omega_t - s_t)\,C_\lambda \right.$$
$$\left. (\times) g\Big(s_t, z(s_t - \rho(s_t)), \int_0^{s_t} F(s_t, \tau_t, z(\tau_t - \rho(\tau_t)))d\tau_t\Big) ds_t\right].$$

Also

$$||Bu(z, \omega_t)||$$
$$= \Big|\Big|W^{-1}\Big[z_1 - T_p(\omega_t)[\varphi(0) + f(z)(0)] - \lim_{\lambda \to +\infty} \int_0^{\omega_t} S_p(\omega_t - s_t)\,C_\lambda$$
$$(\times)\; g\Big(s_t, z(s_t - \rho(s_t)), \int_0^{s_t} F(s_t, \tau_t, z(\tau_t - \rho(\tau_t)))d\tau_t\Big) ds_t\Big]\Big|\Big|$$
$$\leq M_2 M_3\left[||z_1|| - \frac{M_p}{\Gamma(1+p)}||\varphi(0) + f(z)(0)|| - \frac{\sigma M}{\Gamma(p)} \int_0^{\omega_t} (\omega_t - s_t)^{p-1} m_g^*(s_t) ds_t\right]$$
$$\leq M_2 M_3 l_b^* N,$$

where

$$l_b^* = ||z_1|| - ||T_p(\omega_t)\varphi(0)|| - \frac{\sigma M}{\Gamma(p)} \int_0^{\omega_t} (\omega_t - s_t)^{p-1} m_g^*(s_t) ds_t.$$

Hence, the solution becomes

$$\begin{aligned} z(\omega_t) = & T_p(\omega_t)[\varphi(0) + f(z)(0)] \\ & + \lim_{\lambda \to +\infty} \int_0^{\omega_t} S_p(\omega_t - s_t) W^{-1} \Big[z_1 - T_p(s_t)[\varphi(0) + f(z)(0)] \\ & - \lim_{\lambda \to +\infty} \int_0^{s_t} S_p(s_t - \tau_t) C_\lambda g\Big(\tau_t, z(\tau_t - \rho(\tau_t)), \\ & \int_0^{\tau_t} F(\tau_t, \varpi_t, z(\varpi_t - \rho(\varpi_t))) d\varpi_t\Big) d\tau_t \Big] ds_t \\ & + \lim_{\lambda \to +\infty} \int_0^{\omega_t} S_p(\omega_t - s_t) C_\lambda g\Big(s_t, z(s_t - \rho(s_t)), \\ & \int_0^{s_t} F(s_t, \tau_t, z(\tau_t - \rho(\tau_t))) d\tau_t\Big) ds_t. \end{aligned}$$

For convenience, we define $\phi(z(\omega_t)) = \int_0^{\omega_t} F(\omega_t, s_t, z(s_t - \rho(s_t))) ds_t$,

$$\begin{aligned} z(\omega_t) = & T_p(\omega_t)[\varphi(0) + f(z)(0)] \\ & + \lim_{\lambda \to +\infty} \int_0^{\omega_t} S_p(\omega_t - s_t) W^{-1} \Big[z_1 - T_p(s_t)[\varphi(0) + f(z)(0)] \\ & - \lim_{\lambda \to +\infty} \int_0^{s_t} S_p(s_t - \tau_t) C_\lambda g\Big(\tau_t, z(\tau_t - \rho(\tau_t)), \phi(z(\tau_t))\Big) d\tau_t \Big] ds_t \\ & + \lim_{\lambda \to +\infty} \int_0^{\omega_t} S_p(\omega_t - s_t) C_\lambda g\Big(s_t, z(s_t - \rho(s_t)), \phi(z(s_t))\Big) ds_t. \end{aligned}$$

For $\omega_t \in C(J, X_0)$, construct $(\Phi z)(\omega_t) = (\Phi_1 z)(\omega_t) + (\Phi_2 z)(\omega_t)$, where $(\Phi_1 z)(\omega_t) = T_p(\omega_t)[\varphi(0) + f(z)(0)]$ and
$(\Phi_2 z)(\omega_t) = \lim_{\lambda \to +\infty} \int_0^{\omega_t} S_p(\omega_t - s_t) Bu(z, s_t) ds_t + \lim_{\lambda \to +\infty} \int_0^{\omega_t} S_p(\omega_t - s_t) C_\lambda g$
$\Big(s_t, z(s_t - \rho(s_t)), \phi(z(s_t))\Big) ds_t$
for all ω_t in J. Let $B_r(J) = \Big\{\omega_t \in C(J, X_0) : \|\omega_t\| \le r\Big\}$.

Lemma 13.5 *If* $(H1)-(H4)$ *holds,* $\{\Phi z : \omega_t \in B_r(J)\}$ *is equicontinuous.*

Proof. By Lemma 13.3, $T_p(\omega_t)$ is continuous uniformly on J. Subsequently, $\{\Phi_1(z) : \omega_t \in B_r(J)\}$ is equicontinuous. For $\omega_t \in B_r(J)$, assuming $\omega_{t1} = 0, 0 \leq \omega_{t2} \leq b$, we obtain

$$
\begin{aligned}
&|(\Phi_2 z)(\omega_{t2}) - (\Phi_2 z)(0)| \\
&\quad = \lim_{\lambda\to+\infty}\left|\int_0^{\omega_{t2}} S_p(\omega_{t2}-s_t)Bu(z,s_t)ds_t\right. \\
&\qquad \left. + \lim_{\lambda\to+\infty}\int_0^{\omega_{t2}} S_p(\omega_{t2}-s_t)C_\lambda g\Big(s_t, z(s_t-\rho(s_t)), \Phi(z(s_t))\Big)ds_t\right| \\
&\qquad \leq \frac{\sigma M}{\Gamma(p)}\int_0^{\omega_{t2}} (\omega_{t2}-s_t)^{p-1}\Big(l_b^* + m_g^*(s_t)\Big)ds_t \to 0 \text{ as } \omega_{t2}\to 0.
\end{aligned}
$$

For $0 < \omega_{t1} < \omega_{t2} \leq b$,

$$
\begin{aligned}
&|(\Phi_2 z)(\omega_{t2}) - (\Phi_2 z)(\omega_{t1})| \\
&\leq \frac{\sigma M}{\Gamma(p)}\left|\int_{\omega_{t1}}^{\omega_{t2}} (\omega_{t2}-s_t)^{p-1}\Big(M_2M_3l_b^* + m_g^*(s_t)\Big)ds_t\right| \\
&+ \frac{2\sigma M}{\Gamma(p)}\int_0^{\omega_{t1}} \Big[(\omega_{t1}-s_t)^{p-1} - (\omega_{t2}-s_t)^{p-1}\Big]\Big(M_2M_3l_b^* + m_g^*(s_t)\Big)ds_t \\
&+ \left|\lim_{\lambda\to+\infty}\int_0^{\omega_{t1}} (\omega_t - s_t)^{p-1}\left[Q_p(\omega_{t2}-s_t) - Q_p(\omega_{t1}-s_t)\right]\right. \\
&\left.\times C_\lambda g\Big(s_t, z(s_t-\rho(s_t)), \phi(z(s_t))\Big)ds_t\right| \\
&\leq I_1 + I_2 + I_3,
\end{aligned}
$$

where

$$
\begin{aligned}
I_1 &= \frac{\sigma M}{\Gamma(p)}\left|\int_0^{\omega_{t2}} (\omega_{t2}-s_t)^{p-1}\Big(M_2M_3l_b^* + m_g^*(s_t)\Big)ds_t\right. \\
&\quad \left. - \int_0^{\omega_{t1}} (\omega_{t1}-s_t)^{p-1}\Big(M_2M_3l_b^* + m_g^*(s_t)\Big)ds_t\right|, \\
I_2 &= \frac{2\sigma M}{\Gamma(p)}\int_0^{\omega_{t1}} \Big[(\omega_{t1}-s_t)^{p-1} - (\omega_{t2}-s_t)^{p-1}\Big]\Big(M_2M_3l_b^* + m_g^*(s_t)\Big)ds_t,
\end{aligned}
$$

$$I_3 = \left| \lim_{\lambda \to +\infty} \int_0^{\omega_{t1}} (\omega_{t1} - s_t)^{p-1} \left[Q_p (\omega_{t2} - s_t) - Q_p (\omega_{t1} - s_t) \right] \right.$$

$$\left. (\times)\ C_\lambda g\Big(s_t, z (s_t - \rho(s_t)), \phi (z (s_t))\Big) ds_t \right|.$$

By $(H4)$, we can deduce that $\lim_{\omega_{t2} \to \omega_{t1}} I_1 = 0$. Noting that $\left[(\omega_{t1} - s_t)^{p-1} - (\omega_{t2} - s_t)^{p-1}\right]\Big(M_2 M_3 l_b^* + m_g^*(s_t)\Big) \to 0$, as $\omega_{t2} \to \omega_{t1}$, which implies that $\lim_{\omega_{t2} \to \omega_{t1}} I_2 = 0$. For $\epsilon > 0$ small enough, by $(H4)$, we have

$$\begin{aligned}
I_3 \le M & \int_0^{\omega_{t1} - \epsilon} (\omega_{t1} - s_t)^{p-1} \left| Q_p (\omega_{t2} - s_t) - Q_p (\omega_{t1} - s_t) \right| \\
& (\times) \left| g\Big(s_t, z (s_t - \rho(s_t)), \phi (z (s_t))\Big) \right| ds_t \\
& + M \int_{\omega_{t1} - \epsilon}^{\omega_{t1}} (\omega_{t1} - s_t)^{p-1} \left| Q_p (\omega_{t2} - s_t) - Q_p (\omega_{t1} - s_t) \right| \\
& (\times) \left| g\Big(s_t, z (s_t - \rho(s_t)), \phi (z (s_t))\Big) \right| ds_t \\
& \le I_{31} + I_{32} + I_{33},
\end{aligned}$$

where

$$I_{31} = \frac{r\Gamma(p)}{\sigma M} \sup_{s_t \in |\omega_{t1} - \epsilon|} \left| Q_p (\omega_{t2} - s_t) - Q_p (\omega_{t1} - s_t) \right|,$$

$$I_{32} = \frac{\sigma M}{\Gamma(p)} \left| \int_0^{\omega_{t1}} (\omega_{t1} - s_t)^{p-1} m_g^*(s_t)\, ds_t - \int_0^{\omega_{t1} - \epsilon} (\omega_{t1} - \epsilon - s_t)^{p-1} m_g^*(s_t)\, ds_t \right|,$$

$$I_{33} = \frac{\sigma M}{\Gamma(p)} \int_0^{\omega_{t1}} \left| (\omega_{t1} - \epsilon - s_t)^{p-1} - (\omega_{t1} - s_t)^{p-1} \right| m_g^*(s_t)\, ds_t.$$

By Lemma 13.3, it follows that $I_{31} \to 0$ as $\omega_{t2} \to \omega_{t1}$. Applying the arguments, we get $I_{32} \to 0$ and $I_{33} \to 0$ as $\epsilon \to 0$. Thus, $I_3 \to 0$ independently of $z \in B_r(J)$ as $\omega_{t2} \to \omega_{t1}$, $\epsilon \to 0$. Hence, $|(\Phi_2 z)(\omega_{t2}) - (\Phi_2 z)(\omega_{t1})| \to 0$ as $\omega_{t2} \to \omega_{t1}$, which implies $\{\Phi_2 z : z \in B_r(J)\}$ is equicontinuous.

Lemma 13.6 *If* $(H1)-(H4)$ *valid,* $\Phi : B_r(J) \to B_r(J)$ *is continuous.*

Proof. **Claim I:** $\Phi : B_r(J) \to B_r(J)$. For $r > 0$ and by $(H4)$ fulfills

$$M\left(|\varphi(0)| + N + \sup_{\omega_t \in J}\left\{\frac{\sigma M}{\Gamma(p)}\int_0^{\omega_t}(\omega_t - s_t)^{p-1}\Big(M_2M_3l_b^* + m_g^*(s_t)\Big)ds_t\right\}\right) \leq r.$$

For any $z \in B_r(J)$, by Lemma 13.3,

$$|(\Phi z)(\omega_t)| \leq \Big|T_p(\omega_t)[\varphi(0) + f(z)(0)]\Big| + \left|\lim_{\lambda\to+\infty}\int_0^{\omega_t}S_p(\omega_t - s_t)Bu(z, s_t)ds_t\right|$$

$$+ \Big|\lim_{\lambda\to+\infty}\int_0^{\omega_t}S_p(\omega_t - s_t)C_\lambda g\Big(s_t, z(s_t - \rho(s_t)), \phi(z(s_t))\Big)ds_t\Big| \leq r.$$

Hence, $\|\Phi z\| \leq r$, for any $z \in B_r(J)$.

Claim II. Φ is continuous in $B_r(J)$. For some z_m, $z \in B_r(J)$, $m = 1, 2, \ldots$ with $\lim_{m\to\infty} z_m = z$, by $(H3)$, we get

$$g\Big(s_t, z_m(s_t - \rho(s_t)), \phi(z_m(s_t))\Big) \to g\Big(s_t, z(s_t - \rho(s_t)), \phi(z(s_t))\Big) \text{ as } m \to \infty.$$

Using $(H4)$, for each ω_t in J, we get

$$(\omega_t - s_t)^{p-1}\left|g\Big(s_t, z_m(s_t - \rho(s_t)), \phi(z_m(s_t))\Big) - g\Big(s_t, z(s_t - \rho(s_t)), \phi(z(s_t))\Big)\right|$$

$$\leq 2(\omega_t - s_t)^{p-1}m_g^*(s_t),$$

almost everywhere in $[0, \omega_t)$. As $s_t \to 2(\omega_t - s_t)^{p-1}m_g^*(s_t)$ is integrable for $s_t \in [0, \omega_t)$ and $\omega_t \in J$, using Lebesgue theorem,

$$\int_0^{\omega_t}(\omega_t - s_t)^{p-1}\Big|g\Big(s_t, z_m(s_t - \rho(s_t)), \phi(z_m(s_t))\Big) - g\Big(s_t, z(s_t - \rho(s_t)), \phi(z(s_t))\Big)\Big|ds_t$$

$\to 0$ as $m \to \infty$. For $\omega_t \in J$, we assures

$$\Big|\Big(\Phi z_m\Big)(\omega_t) - (\Phi z)(\omega_t)\Big| \leq \frac{\sigma M}{\Gamma(p)}\int_0^{\omega_t}(\omega_t - s_t)^{p-1}\Big|g\Big(s_t, z_m(s_t - \rho(s_t)), \phi(z_m(s_t))\Big) - g\Big(s_t, z(s_t - \rho(s_t)), \phi(z(s_t))\Big)\Big|ds_t \to 0 \text{ as } m \to \infty.$$

Therefore, $(\Phi z_m) \to (\Phi z)$ pointwise on J as $m \to \infty$. By Theorem 13.1, (Φz_m) uniformly converges to (Φz) on J as $m \to \infty$. Hence, Φ is continuous.

Theorem 13.2 *If* $(H1)-(H9)$ *are valid, the non-local system* (13.1)–(13.2) *is controllable on* $[0, b]$ *provided* Φ *is continuous and*

$$\frac{2\sigma M l_g^* l_f^* \nu^* \zeta^*}{\Gamma(p)} \int_0^{\omega_t} (\omega_t - s_t)^{1-p} \alpha(\{\zeta(s_t)\})\, ds_t \le r.$$

Proof. We have $\Phi : B_r \to B_r$ is continuous. Assume $\zeta \subset B_r$ is countable $\zeta \subset \overline{co}(\{0\} \cup \Phi(\zeta))$, then we prove $\Phi(\zeta) = 0$.

Let $V_0(\omega_t) = T_p(\omega_t)[\varphi(0) + f(z)(0)]$, for all ω_t in J and $V_{m+1} = \Phi V_m$, $m = 0, 1, 2, \ldots$. Define $\zeta = \{V_m : m = 0, 1, 2, \ldots\}$. By Lemma 13.5, ζ is equicontinuous and uniformly bounded on J. Ensure $\zeta = \{V_m : m = 0, 1, 2, \ldots\}$ is relatively compact in X_0, for any ω_t in J. From $(H5)$, $(H6)$ and Lemma 2.3, for $\omega_t \in J$, we get

$$\begin{aligned}\alpha(\zeta(\omega_t)) = \alpha(\{V_m(\omega_t)\}_0^\infty) &= \alpha(\{V_0(\omega_t)\} \cup \{V_m(\omega_t)\}_{m=1}^\infty) \\ &= \alpha(\{V_m(\omega_t)\}_{m=1}^\infty)\end{aligned}$$

and

$$\begin{aligned}&\alpha(\{V_m(\omega_t)\}_{m=1}^\infty) \\ &\le \frac{2\sigma M}{\Gamma(p)} \int_0^{\omega_t} (\omega_t - s_t)^{p-1} \alpha(g(s_t, \{V_m(\omega_t - \rho(\omega_t))\}_{m=0}^\infty, \phi(\{V_m(\omega_t)\}_{m=1}^\infty)))\, ds_t.\end{aligned}$$

Also

$$\alpha(\{V_m(\omega_t)\}_{m=1}^\infty) \le \frac{2\sigma M}{\Gamma(p)} \int_0^{\omega_t} (\omega_t - s_t)^{p-1} \alpha\Big(Bu(\{V_m(\omega_t)\}_{m=0}^\infty, s_t)\Big)\, ds_t.$$

Thus

$$\alpha(\{\zeta(\omega_t)\}) \le \frac{2\sigma M}{\Gamma(p)} l_g^* l_f^* \nu^* \zeta^* \int_0^{\omega_t} (\omega_t - s_t)^{1-p} \alpha(\{\zeta(s_t)\})\, ds_t.$$

Using Gronwall's inequality, we conclude $\alpha(\{\zeta(\omega_t)\}) = 0$. Followed by, $\zeta(\omega_t)$ is relatively compact. By Lemma 13.4, Φ has at least one fixed point z in B_r such that $z(b) = z_1$. Hence, the system (13.1)–(13.2) is controllable.

13.4 Conclusion

This article illustrates that the controllability results on fractional integro-differential system with delay behaviour is non-densely defined on Banach space using the fractional calculus, measure of non-compactness and Monch fixed point theorem. Our theorem guarantees the effectiveness of controllability, which is the result of the system concerned. It would be interesting if we investigate the same in Hilbert space by employing some other fixed point theorems that are suitable to the nature of the system. Moreover, we can explore the optimal controllability for various types of fractional integro-differential equations. It has many significant applications not only in control theory and systems theory but also in such fields as industrial and chemical process control, reactor control and control of electric bulk power systems, medical sciences, aerospace engineering and recently in quantum systems theory. In particular, the non-linear fractional integro-differential equation is implemented for the ECG images to detect the abnormal heart rates of the patient.

References

[1] Abada, N., Benchohra, M., and Hammouche, H. (2009), Existence and controllability results for nondensely defined impulsive semilinear functional differential inclusions, *Journal of Differential Equations*, **246(10)**, 3834-3863.

[2] Agarwal, R. P., Dos Santos, J. P. C., and Cuevas, C. (2012), Analytic resolvent operator and existence results for fractional integrodifferential equations, *Journal of Abstract Differential Equations and Applications*, **2**, 26–47.

[3] Bahuguna, K. D. (2016), Approximate controllability of nonlocal neutral fractional integrodifferential equation with finite delay, *Journal of Dynamical and Control Systems*, **22(3)**, 485–504.

[4] Banas, J. and Goebel, K. (1980), Measure of noncompactness in Banach spaces, *Lecture Notes in Pure and Applied Mathematics*, **60**, Marcel Dekker, New York.

[5] Da Prato, G. and Sinestrari, E. (1987), Differential operators with nondense domain, *Annali Della Scuola Normale Superiore Di Pisa-Classe Di Scienze*, **14**, 285–344.

[6] Fu, X. (2004), On solutions of neutral nonlocal evolution equations with nondense domain, *Journal of Mathematical Analysis and Applications*, **299**, 392–410.

[7] Fu, X. and Liu, X. (2007), Controllability of nondensely defined neutral functional differential systems in abstract space, *Chinese Annals of Mathematics*, **28(2)**, 243–252.

[8] Gatsori, E. P. (2004), Controllability results for nondensely defined evolution differential inclusions with nonlocal conditions, *Journal of Mathematical Analysis and Applications*, **297**, 194–211.

[9] Goswami, A., Singh, J., Kumar, D., and Gupta, S. (2019), An efficient analytical technique for fractional partial differential equations occurring in ion acoustic waves in plasma, *Journal of Ocean Engineering and Science*, doi:10.1016/j.joes.2019.01.003.

[10] Gu, H., Zhou, Y., Ahmad, B., and Alsaedi, A. (2017), Integral solutions of fractional evolution equations with nondense domain, *Electronic Journal of Differential Equations*, **145**, 1–15.

[11] Henderson, J. and Ouahab, A. (2007), Controllability of nondensely defined impulsive functional semilinear differential inclusions in Frechet spaces, *International Journal of Applied Mathematics and Statistics*, **9**, 35–54.

[12] Hilfer, R. (2000), *Applications of Fractional Calculus in Physics,* World Scientific, Singapore.

[13] Jothimani, K., Valliammal, N., and Ravichandran, C. (2018), Existence result for a neutral fractional integrodifferential equation with state dependent delay, *Journal of Applied Nonlinear Dynamics*, **7(4)**, 371–381.

[14] Kavitha, V. and Mallika Arjunan, M. (2010), Controllability of nondensely defined impulsive neutral functional differential systems with infinite delay in Banach spaces, *Nonlinear Analysis: Hybrid Systems*, **4**, 441–450.

[15] Kilbas, A. A., Srivastava, H. M., and Trujillo, J. J. (2006), Theory and applications of fractional differential equations In: North-Holland Mathematics Studies, **204**, Elsevier Science, Amsterdam.

[16] Kucche, K. D., Chang, Y. K., and Ravichandran, C. (2016), Results on non-densely defined impulsive volterra functional integrodifferential equations with infinite delay, *Nonlinear Studies*, **23(4)**, 651–664.

[17] Mahmudov, N. I., Murugesu, R., Ravichandran, C., and Vijayakumar, V. (2017), Approximate controllability results for fractional semilinear integro-differential inclusions in Hilbert spaces, *Results in Mathematics*, **71 (1–2)**, 45–61.

[18] Mokkedem, F. and Fu, X. (2014), Approximate controllability of semilinear neutral integrodifferential systems with finite delay, *Applied Mathematics and Computation*, **242**, 202–215.

[19] Mophou, G. M. and N'Guerekata, G. M. (2009), On integral solutions of some nonlocal fractional differential equations with nondense domain, *Nonlinear Analysis*, **71**, 4665–4675.

[20] Lakshmikantham, V., Leela, S., and Vasundhara Devi, J. (2009), *Theory of Fractional Dynamic Systems*, Cambridge Scientific Publishers, Cambridge.

[21] Pazy, A. (1983), *Semigroups of Linear Operators and Applications to Partial Differential Equations*, Springer-verlag, New York.

[22] Podlubny, I. (1999), *Fractional Differential Equations. An Introduction to Fractional Derivatives, Fractional Differential Equations, to Methods of Their Solution and Some of Their Applications*, Academic Press, San Diego, CA.

[23] Ravichandran, C. and Baleanu, D. (2013), Existence results for fractional neutral functional integrodifferential evolution equations with infinite delay in Banach spaces, *Advances in Difference Equations*, **1**, 215–227.

[24] Ravichandran, C., Jothimani, K., Baskonus, H. M., and Valliammal, N. (2018), New results on nondensely characterized integrodifferential equations with fractional-order, *The European Physical Journal Plus*, **133(109)**, 1–10.

[25] Ravichandran, C., Valliammal, N., and Nieto, J. J. (2019), New results on exact controllability of a class of neutral integrodifferential systems with state dependent delay in Banach spaces, *Journal of the Franklin Institute*, **356(3)**, 1535–1565.

[26] Suganya, S., Kalamani, P., and Mallika Arjunan, M. (2016), Existence of a class of fractional neutral integrodifferential systems with state dependent delay in Banach spaces, *Computers and Mathematics with Applications*, 1–17.

[27] Subashini, R., Vimal Kumar, S., Saranya, S., and Ravichandran, C. (2018), On the controllability of non-densely defined fractional neutral functional differential equations in Banach spaces, *International Journal of Pure and Applied Mathematics*, **118(11)**, 257–276.

[28] Thieme, H. R. (1990), Integrated semigroups and integrated solutions to abstract Cauchy problems, *Journal of Mathematical Analysis and Applications*, **152**, 416–447.

[29] Valliammal, N. and Ravichandran, C. (2018), Results on fractional neutral integrodifferential systems with state dependent delay in Banach spaces, *Nonlinear Studies*, **25(1)**, 159–171.

[30] Valliammal, N., Ravichandran, C., and Park, J. H. (2017), On the controllability of fractional neutral integrodifferential delay equations with nonlocal conditions, *Mathematical Methods in the Applied Sciences*, **40(14)**, 5044–5055.

[31] Prakash Dubey, V., Kumar, R., and Kumar, D. (2019), Analytical study of fractional Bratu-type equation arising in electro-spun organic nanofibers elaboration, *Physica A: Statistical Mechanics and its Applications*, **521**, 762–772.

[32] Vijayakumar, V., Ravichandran, C., Murugesu, R., and Trujillo, J. J. (2014), Controllability results for a class of fractional semilinear integro-differential inclusions via resolvent operators, *Applied Mathematics and Computation*, **247**, 152–161.

[33] Vijayakumar, V., Ravichandran, C., and Murugesu, R. (2013), Nonlocal controllability of mixed Volterra-Fredholm type fractional semilinear integrodifferential inclusions in Banach spaces, *Dynamics of Continuous, Discrete and Impulsive Systems Series B: Applications & Algorithms*, **20**, 485–502.

[34] Vijayakumar, V. (2018), Approximate controllability results for non-densely defined fractional neutral differential inclusions with Hille Yosida operators, *International Journal of Control*, **92(9)**, 1–13.

[35] Wang, W. and Zhou, Y. (2012), Complete controllability of fractional evolution systems, *Communications in Nonlinear Science and Numerical Simulation*, **17(11)**, 4346–4355.

[36] Zhou, Y. and Jiao, F. (2010), Nonlocal Cauchy problem for fractional evolution equations, *Nonlinear Analysis: Real World Applications*, **11**, 4465–4475.

[37] Zhou, Y. (2014), *Basic Theory of Fractional Differential Equations*, World Scientific, Singapore.

Index